Generis

PUBLISHING

The Oral Epic in the Western World

Studies in Applied Fundamental, Structural, and
Generative Anthropology

Robert Rois

Title: The Oral Epic in the Western World

Studies in Applied Fundamental, Structural, and Generative Anthropology

ISBN: 979-8-88676-539-7

Author: Robert Rois

Cover image: www.pixabay.com

Publisher: Generis Publishing
Online orders: www.generis-publishing.com
Contact email: info@generis-publishing.com

Table of Contents

To the memory of Elroy L. Bundy, professor and friend

Acknowledgement

The two essays on *The Song of Roland* appeared first in *Anthropoetics* a few years back. The *Review of European Studies* printed the *Mio Cid* paper in 2021. More recently, the *Global Journal of Human Social Sciences* published our *Beowulf* monograph. *Anthropoetics* has just published the *Iliad* paper. They all graciously allow these works to be included in the present volume.

Foreword

I admire my former student Robert Rois' tenacity in remaining loyal to his scholarly roots, and welcome this volume of studies of the ancient and medieval epic as a worthy contribution to scholarship in his field. Although I am not fully qualified to judge his contributions to the four areas of Homeric, Beowulf, Roland, and Cid studies, I would like to suggest why I believe specialists in these fields can find valuable insights in these essays.

Robert's examination of what might be called "relationship clusters" in the *Iliad* provides an original insight into the "deep structure" of the poem. How does the origin of the poem in the resentful "wrath" of Achilles over Agamemnon's arbitrary appropriation of his captive Briseis play out in the context of his friendship with Patroclus, who befriends her, and his rivalry with Hector, who similarly befriends Helen? And how does the overall pattern of their interactions, ending with Achilles' return of Hector's body to Priam and the subsequent funeral celebration, reflect the passage from a world of military rivalry to one of contractual exchange?

The *Chanson de Roland* is the masterpiece of the French *chanson de geste*. Robert's two essays on the poem are not intended to provide an overall interpretation of this work, but rather point to areas whose exploration might encourage the specialist to reexamine it from new perspectives. How does the *ostensive* as opposed to *imperative* gesture of sounding the oliphant transform it into an act of Christian faith that raises the chivalric sense of honor to the level of genuine martyrdom? How does the understanding of the complex family connections of Charlemagne, Ganelon, and Roland with the (unmentioned) Dame Gile permit us to explain the *padrastre*'s hostility that ironically enables Roland's martyrdom—contributing to the status of the *Chanson* as the greatest of all French *chansons de geste*?

Similar questions are raised by the other essays, whether by the boast/promise duality in *Beowulf* or the Cid's preference of moral to physical triumph over his high-born enemies. All five essays introduce new perspectives, grounded on close textual readings in the original languages, into the study of the masterworks of European epic. Specialists in the domain of Homeric and medieval studies should welcome the challenge posed by these five probing analyses.

Eric Gans
Distinguished Professor of French, Emeritus
University of California, Los Angeles

Introduction

Often we wish to understand the origin of our Western Culture. Although to reconstruct history from the past is difficult, we yet possess record of the tradition which existed in Classical Antiquity and then again from the eighth through the thirteenth centuries. The best known oral Western Epics contain the customs, beliefs, and language of civilization and maintain traditions predominant in centuries before and after.

Through conceptual semantic analysis of the *Iliad* and the main oral epics in Medieval Europe we can expose the important ideals that influenced the way our civilization developed. We depart from the premise that the oral poets plied their trade in Greece, Britain, France and Spain divulging customs that unfold in complex tales of heroes battling foes, always in order to derive codes of behavior and moral standards acceptable as entertainment, long before the advent of written record. The actual extant manuscripts date from the ninth through the thirteenth century on; but the tales cover a period extending over centuries before. Scribes in monasteries copied the texts that eventually reached us. Milman Parry and Albert Lord have delineated standard procedures used in forging the long oral poems we consider as central to the development of Western Literature. [1] We pose as plausible the theory that since the Classical Age a common tradition was passed on by oral poets, the *rhapsodes* of ancient Greece, who we know collectively as *Homer*. The author responsible for composing the *Iliad* and the *Odyssey* is said to have been blind. This tradition probably stems from the possibility that Homer, and his peers, could not read a language which may not yet have become burdened by a definitive script. The fact that the Homeric epic contains evidence of oral composition discloses the fact that it was originally derived orally. Who could possibly take down in literate form a popular oral tradition forged by perhaps

---

[1] Foley 19.

illiterate poets? There were no stenographers. The scribes gathered manuscripts centuries later. Common sense dictates the incontrovertible supposition that a *rhapsode* became in time literate enough to take down his own performance. Record of this brave, original effort by the hypothetical "last rhapsode" is not extant. Later manuscripts copied over and over, and handed down through generations gradually came down to us.

The extant copies handed down to us through the centuries reveal that the original tales of heroes, gods, monsters, wars, and human toil in general, disclose the beliefs, and customs commonly referred to as shared mythology. The people's religion was important. Collectively, the tales divulged the norms and standards that became a source for the legal precepts and religious beliefs that were to be cherished by posterity.

Through the fourth to the ninth centuries, the epoch known as the Dark Ages, or preferably, as Late Classical Antiquity, newer considerations for beliefs and social customs sprang up throughout Europe. When we deal with the Medieval Oral Epic we touch upon a wealth of material which allows us to ponder over the original development for the modern means of symbolic representation which evolved from the actual roots. In the language of the Oral Epic we find the foundation for our socio-cultural heritage. The Homeric Epic appeared at the advent of Western literature; likewise the Medieval Oral Epic sprang as the vernacular languages started to define themselves from an older, more established linguistic source. We import our hypothesis about the last *rhapsode* on to the Middle Ages. The last *scop* was the one who fortunately became literate enough and took down in the West Saxon dialect the text we know as *Beowulf*, a product from centuries of oral composition. We can be sure that this original manuscript is lost to the Ages. The same could be said of the last *jongleur*, who took down the *Roland* epic, and the last *juglar* who preserved in written form the *Mio Cid* poem. We value the scholarly editions compiled by dedicated individualists such as Colin Smith, and Joseph Bédier; their trust on a literate author has made it possible for us to rely on solid documents for our reading and understanding. Yet

14

we must insist that relevance for the study of the Oral Epic must be established based on the belief expounded by traditionalists such as Ramón Menéndez Pidal.[2] We are dealing with documents not product of a single author. Our responsibility in these essays is to decipher the origin of cultural norms, legal thought, and religious principles as they are revealed by countless individuals through syntactic and semantic structures in a linguistic construct used before written literature. The tales of heroes reveal the development of a sprouting civilization. By the tenth century the vernacular languages, French and Spanish from Latin, and English from Old Norse, had achieved written consistency. Reading *Beowulf*, *The Song of Roland*, and *Mio Cid* we dwell in the malleable expressive texture of a new age. For this reason we can not take for granted these precious literary documents; they represent a primordial linguistic framework for the socio-cultural foundation of our Western civilization.

Our treatment of the *Iliad* depends on René Girard's theory of the mimetic double bind, which could be extended from the typical love triangle on to a quadrangular relationship, to unravel events in the plot that show how a war poem can have a peaceful ending. In the case of *Beowulf* we trace the development of political, religious, and legal principles through language. Our conceptual semantic analysis for the rhetorical use of conditional statements relies on Claude Lévi-Strauss' anthropological classification of opposing binary structures in primitive societies. Consistently, hypothetical statements, or "if" clauses, enclose positive and negative ramifications. A boast at the mead hall can become a challenge when facing the enemy; a prayer for victory could become a curse to an antagonist. The formulation of an oath opens up possibility for breach and betrayal. Then we go on to explore the notion of *exception* in order to derive implications in legal developmental history.

---

[2] Foley 76. John Miles Foley explains how Joseph J. Duggan proposed a threshold of 20 percent straight formulas "as the minimum measure for the orality of a poem." Foley 79-80. He found *Mio Cid* to achieve 31.7 percent, and *Roland* 35 percent, both above the 20 percent suggested threshold. Foley 80.

When we examine *The Song of Roland* we diligently decipher the possible meaning for a non verbal sound, the *Oliphant*. Delaying the call for help is baffling. In this case we deal with notions for semantic considerations expounded by Eric Gans through his work in Generative Anthropology. Once we have established the prevalence of one side over another, we then turn to reasonable implications for the guilt that overwhelms the main characters along development of the story. René Girard's mimetic definition of desire provides a framework for explaining the cause for Ganelon's treason in the *Roland* epic.

We close with an exposé of the dilemma confronted by *Mio Cid* in the Spanish epic. The hero accepts the banishment imposed by the king, defends lands on behalf of the crown, and conquers territory for the monarch. He even withstands the treason of corrupt nobles who beat his daughters in order to establish the conceptual principles that form a basis for moral worth. The scene of banishment, and the unspeakable abuse against his family, collide with the conquering Cid's moral stance and create a complex dimension for his character; the resulting dramatic tension presents a relevant contrast between the historical legend and the hero of the poem. We hope to consolidate this dichotomy.

Throughout all our studies in this volume the notion of victory becomes subsumed under the concept that glory can only be attained through sacrifice. Heroism in this context is redefined in each epic to reach an adequate perspective for viewing a protagonist whose prowess is subservient to the common good. We should bear in mind that popular beliefs are to be seen as occurring within a warring society. The common people in an epic context are the heroes' fellow warriors, and their society consists of members in their armies. From direct confrontation with conflict and strife we may derive a sense of what seems fair and just, distilling an implied concept of essential morality. But there is a limit to how far we can go in identifying the heroic *ethos* with our own idea of that which appears good or evil. Our task is to explore the content illustrated in the *Iliad*, *Beowulf*, *Roland* and *Mio Cid*, and explain how the poems reflect values of their epoch and social milieu using our modern methods of literary analysis.

16

Victory through Defeat

When seeking proper interpretation for the oral tales of great heroes, we should ponder over what could possibly be conceived as suitable reward for confronting death. In order to understand where glory resides we must relinquish any notion of material reward and the actual success which may block our concern for posterity while seeking instant victory. The real hero in the Oral Epic does exactly what he must regardless of the immediate outcome. On the figures of such heroes as Achilles, Beowulf, Roland and Mio Cid rest views regarding the future of civilization. These heroes face defeat in order to fulfill their mission while attaining the paradigmatic roles we greatly admire; paradoxically, they are models we cannot emulate without assuming a spirit of sacrifice. The hero's sacrificial stance becomes the lesson of the great Oral Western Epics.

We choose for our study the Oral Epic as a specific genre, excluding all else. In his *From Virgil to Milton* C.M. Bowra reminds us of the known distinction between what is called the *authentic epic*, as opposed to the *learned epic*.[1] The most obvious distinction is that one springs from popular belief and lore, a tradition that extends over centuries; the other is product of one author. The moment when epic emerges from oral tales takes place within oral culture. Relevance for our study of the Oral Epic depends on this distinction. The *Iliad* ends with the burial of Hector whose corpse was ransomed by Priam. The Trojan King approached his son's slayer, Achilles, who ceases to drag around Patroclus' tomb the body of Hector; the hero turns the body over to Priam and shares a meal with the King of Troy. We know that Achilles did not heed Hector's plea for proper burial of the vanquished during their duel, but their combat is not the end of the poem. Consequently, regardless of the actual cruelty of true battle, *The*

---

[1] Bowra 1.

Iliad, a Classical authentic Oral Epic, closes on a forgiving note. Such implied acceptance of sacrifice is the kernel for resolution of the conflict. Submission to the peaceful entreaties of Priam is a sacrifice for Achilles, nearly as great as Roland's sacrificial death. [2] The anger and suffering of Achilles' at Patroclus' loss parallels Charles' bitterness at Roland's death in *The Song of Roland*, where, at the end, execution of Ganelon brings a legal closure. Ransoming of Hector's corpse assures that a peaceful contractual agreement ends the *Iliad*, revealing how Achilles finally subdues his wrath, and allows Priam to mourn his dead son.

While comparing the *Song of Roland* to the *Iliad* we note that, although Ganelon's treason due to resentment parallels Achilles' withdrawal through anger, the Greek warrior is no traitor. In contrasting plot overview between the *Iliad* and other oral epics from the Middle Ages we note that individual differences are superseded by general similarities. The internal struggle centered on the grieving hero's overbearing anger over his lost mistress is resolved: a) by Achilles' acceptance of the restitution formerly refused, b) the warrior's return to the fighting after the loss of Patroclus, and c) the furious hero's return of Hector's corpse to Priam at the end of the epic. On the other hand, the *Roland* strife, completely integrated within the family unit, acquires a deeper moral configuration leading to a bitterly tragic family drama in the Age of Faith: a) victim of treason, the emperor's nephew dies, b) Charlemagne mourns, and c) Ganelon, the king's brother in law and the hero's stepfather, is executed. Destiny is brutal. The common parallel between the *Iliad* and *The Song of Roland* is that

---

[2] Eric Voegelin interprets the prediction that rules Achilles' destiny as a distinguishing factor that sets him apart from the other Greek leaders: "he lives truly in battle, and his sulking wrath is most painful to maintain while joyous slaughter goes on without him." Voegelin insists: "the other princes are bound to their station by their oath and duty; they cannot return as long as victory has not become obviously hopeless in military terms. Achilles is bound to the war, and can never return because he is a warrior." Voegelin 89. Here we have the thematic foundation for Achilles' resentment. As Voegelin suggests, imperishable fame occurs after death; according to Achilles' nymph mother Thetis' prediction, glory cannot be experienced in life, for it is a reward of the spirit. Voegelin 90-91. By waiting for the precise instance before total disaster overcomes the Greeks to return to battle, so that they acknowledge his worth as a great hero, Achilles loses Patroclus. He must accept the fact that he is to die young with glory, and forgo the joys of old age.

in both oral epics glory and redemption is achieved through loss and sacrifice. Triumph is moderate since victory is attained through defeat. The slaying of Hector avenges but does not relieve the loss of Patroclus, just as the execution of Ganelon cannot make up for the death of Roland. Since the brunt of human loss is total and absolute, the narrative becomes a tale of retribution due to guilt which cannot be entirely appeased, what Eric Gans calls resentment.[3] We are left with the sole resignation brought about by sacrificial redemption. As readers, we participate in the unique cathartic consolation of epic narrative.

Such balanced sacrificial loss is paralleled elsewhere in the oral tradition across border and linguistic frontiers. We may cast a compassionate look at Beowulf, in the epic that bears his name. The hero dies at the end. It is up to the reader to find sublimation and significance in the hero's demise. Roland dies early enough in the course of the *Chanson* to leave us determined to seek a justification for his loss. The trial at the end of *The Song of Roland* seeks to establish a rational response for the hero's death. Mio Cid does not die in the middle, but at the end of the famous *Poema*; during life he endures banishment and the outrage perpetrated against his daughters. The hero endures with resignation. Resolution is brought about by the chastisement of his foes, which the Cid does not witness, preferring to leave the scene of retribution, before the duels commence at the end of the trial. We must seek satisfaction and release in various contexts.

In our polymorphous collection of essays we suggest that victory, or the attainment of fame and glory, is not as evident as general belief may assume it to be. The ordeal of the hero in the Oral Epic is authentic, to apply further significance to the term. The reader, or the audience in the original public rendition of the poems, is left to draw some meaningful explanation regarding the actual nature of the hero's real triumph while confronting his suffering. In each case we seek differentiation between friend and foe. Regardless of whether classification into factions by corresponding clans seems too rigorous or not, we

---

[3] Gans 242, 244, 246.

agree with Emile Durkheim that the *genus* is distinct from the *species*. [4] To which tribe we belong may ultimately determine who we are. We conclude, along with Durkheim, that, since frontiers define nationalistic tendencies, borders must be carefully delineated and protected. [5] Differentiation is crucial to preserve the social order. Especially during time of conquest, questions of good and evil, or even open ego-centricity, cannot contaminate prospects for military defense and geographic consolidation. Since the Oral Epic sprouts at the advent of cultural norms, geographic frontiers, and linguistic expression, to achieve a sensibly accurate reading for each epic poem we should discover through conceptual semantic analysis an interpretation based on widespread anthropological custom and acknowledged common justice.

The branch of studies which we call cultural anthropology adheres to the belief that civilization in its development follows a definite set of principles, or patterns. In this volume we explore the structural anthropology of Claude Lévi-Strauss, the originary thinking of Eric Gans, and the mimetic theory of René Girard. These three thinkers have illustrated the central cultural aspects shared by all humanity which are expressed in language, crossing geographic and ethnic borders. Their work sets down the groundwork for consistent critical interpretation of the Oral Epic. The advent of culture is most clearly expressed by oral tradition in the early stages of language formation.

Our work could not develop without the important premise that, in order to foment cross cultural comparative studies, we must not lose track of the relevant connections. We do not force the subject of research on a particular method of analysis; rather, the method of analysis is adapted to the subject under research. For this reason, we note that the treatment of *Beowulf* we expound could not be

---

[4] Durkheim 47. *Genus* is the main subdivision of a family. A *genus* may include one or more *species*. The *genus* is usually capitalized and precedes the *species*; such is the case with *Homo sapiens*. Webster 1, 765. There are other *species* of *Homo*, the advanced primates. Yet the *genus* can become subordinate to a larger classification. *Man* is a *species* under *animal*, a *genus*; but *man* may be regarded as a *genus* with respect to other *species*, European, Asiatic, etc. Webster 2, 1741.

[5] Durkheim 57.

possible without the opposing binary structures Claude Lévi-Strauss outlines in his *Structural Anthropology*. We rely on Eric Gan's work to show how the semantic basis for the *ostensive* at the origin of language is seen in the strict nature of the dilemma faced by Roland in not sounding the Oliphant until it is too late to avoid tragic death. Our exploration requires understanding of the *ostensive* as emanating from the originary scene, an analysis Gans provides in his *The Origin of Language*, flagship work for Generative Anthropology. Renée Girard's fundamental anthropology, which he later calls mimetic theory, takes us from *Homer* to *Roland*. Just as Helen is an object of desire between Menelaus and Paris, with Hector included as benefactor, the Emperor's sister, Dame Giles, becomes a placeholder in the quadrangular relationship which includes Roland and Ganelon, as mother and wife, respectively. Girard's theory of the mimetic double bind provides us a way to achieve a new understanding for the *Roland* epic. His works *Violence and the Sacred*, and *The Scapegoat* provide ways to view character and plot formation in the *Iliad*, the *Song of Roland*, and the Spanish epic of *Mio Cid*. Not unlike the young hero Achilles, the mature Cid is burdened by the roles of victim and heroic warrior simultaneously. Ambivalence is resolved once a consistent interpretation is achieved.

The influence of *Homer* on the medieval epics of *Beowulf*, *Roland* and *Mio Cid* unites these works in literary history. There is a strong sense of legal closure in the Oral Epic. The *Iliad* ends with the ransoming of Hector's corpse. The *Roland* and *Mio Cid* epics both end with legal proceedings that lead to retribution. The ending of *Beowulf* adds bewilderment to the impact of loss while confronting material wealth in a transitory life. Henceforth, after reading the studies contained in this volume, we should be aware that the Oral Epic constitutes a genre which allows anthropological understanding by virtue of its origin at the advent of literature. Epic narrative emerges gradually from the oral tales which demonstrate the rise of civilization from the traditional customs prevalent in our Western World.

Works Cited

Bowra, C.M. *From Virgil to Milton*. The MacMillan Co. Ltd., 1963.

Durkheim, E. and Mauss, M. *Primitive Classification*. Translated by Rodney Needham. U of Chicago, 1967.

Foley, John Miles. *The Theory of Oral Composition. History and Methodology*. Indiana U P. 1988.

Gans, Eric. *The End of Culture*. U of California, 1985.

Voegelin, Eric. *Order in History*. Vol. 2. *The World of the Polis*. Louisiana State U,

Webster, Noah. *Webster's New Twentieth Century Dictionary of the English Language*. Edited by Jean L. McHechnie. Vols. 1-2. The World Publishing Co., 1965.

Ransom for Desire in the *Iliad*: Hector and Patroclus

In the text of the *Iliad*, we see that the Achaeans are fighting Troy for the return of Helen; yet there are contingencies to be resolved. The main theme of the *Iliad* is the wrath of Achilles and all its dire consequences, although the abduction of Helen is a backdrop which accounts for the clash of two military factions. The Trojan War rages on for nine years, all because Paris, by abducting Helen, had broken oaths of guest to his host Menelaus. But then, during the siege of Troy, at the Greek camp a pestilence breaks out. Achilles calls forth an assembly. The seer Kalchas assigns cause for the plague to Agamemnon's taking of the maid Chryseis from her father Chrises, Apollo's priest. To appease the god and prevent the contagious illness from spreading further in the ranks, Agamemnon must give back Chryseis.[1] Agamemnon accedes but takes away Briseis from Achilles. Subsequently, Achilles withdraws from the fighting to sulk in his tent for the loss of Briseis. Absence of the best warrior occasions countless Greek casualties in the furious fighting. The question arises as to how much grief is justified by the loss of Achilles' female companion. We must make Homer's concerns our own.

Patroclus and Hector share much of the spotlight as central characters in the *Iliad's* plot, for they are both closely involved in the two love triangles which outline the theme of sacrifice in the Homeric epic. The relationship between Achilles and Agamemnon, with Briseis as object of desire, reflects in microcosm the broader upheaval brought about by the struggle between Menelaus and Paris over the abducted Helen, genesis of the Trojan war. In his *The End of Culture* Eric Gans views the distinction between both conflicts as vital source for the epic genre:

---

[1] Werner Jaeger in his *Paideia: The Ideals of Greek Culture*, Vol 1, explains: "The leitmotiv of *Ate* is heard even in this, the first episode of the poem. Agamemnon is infatuated when he commits the first offense, and in Book IX Achilles is blinded by *Ate*. He 'knows not how to yield;' but clings doggedly to his anger and thus exceeds the limit allowed to mortal men." Jaeger 48.

"The fact that Achilles' grievance against Agammemnon mirrors within the Greek camp that of Menelaus against Paris with which the Trojan war began is not a mere sign of what a New Critic might call the 'well wroughtness' of the poem, or an illustration of what French formalists call *mise en abyme*. This involution of the external conflict is precisely the moment of literary epic, as opposed to the literary legend from which it sprang." [2]

The critic incisively suggests that Agamemnon's overbearing role vis à vis Achilles narrowly mirrors Paris' violation of the alliance due to his host, Menelaus, while a guest at his palace. The epic springs from the legend. To maintain the role of victors in the struggle, Argives must differentiate themselves from their Trojan foes. There is disorder in their ranks analogous to the war they wage against Troy. Quarrel over two female characters stand as background for the violent devastation of war.

Helen appears most prominently in the Third Book of the *Iliad*. The abducted heroine comes to the fore as Priam, contemplating the fighting from the walls of Troy, inquires about the fighting heroes. The Trojan King reassures Helen of her innocence in the slaughter by blaming the gods for their ordeal [*Il.* III, 164].[3] Assigning guilt to the gods is an attempt to disclaim responsibility and to discount possibility for human frailty and the ravages exacted by desire. The inhabitants of Troy understand that Helen's beauty may justify her abduction, but not the war. They wish that she would go away on the ships back to Greece, or else she shall become a bane, *pema*, to them and their children [*Il.* III, 156-160].[4] Hector himself rebukes his brother for taking away from her husband a fair woman of a remote land [*Il.* III, 39, 48-49]. Paris justifies the abduction of Helen by reference to her ravishing beauty, adding that gifts from the goddess of love are not to be cast aside

---

[2] Culture 243.

[3] οὔ τί μοι αἰτίη ἐσσί, θεοί νύ μοι αἴτιοί εἰσιν. "You are not guilty to me; the gods now are guilty to me" [*Il.* III, 164]. Trans. mine. Quotes from the original text are from the same URL throughout.

[4] Voegelin 96.

[*Il*. III, 65]. Hector's exoneration of Paris shows his allegiance to Aphrodite, protector of Troy.

At this critical junction in the Third Book of the *Iliad* the two central parties in the conflict arrange for a trial by combat. The winner shall take Helen, bringing a peaceful end to the war and saving lives. But the duel between Menelaus and Paris ends when Helen's lawful husband is left holding Paris' helmet while the legendary womanizer is transported by Aphrodite to the lovely lady's bedchamber. Helen, meanwhile, is still dwelling on her heritage and pointing out the individual prowess of the Greek heroes, standing close to Priam on the Trojan walls. When Aphrodite beckons her to return to the bedroom, where Paris is waiting, Helen rebukes the goddess for such sudden summons. In the battlefield the heroes look for Paris in vain. Understandably, Helen refuses to do "the shameful act," νεμεσσητὸν [*Il*. III, 410]. [5] We are led to believe that Paris is very different in character from the fighting men. But the goddess remains undaunted and threatens Helen the goddess shall throw her to the mercy of both Greeks and Trojans, to be torn to pieces by the rage ensuing from their bitter hatred if she does not comply with Aphrodite's dream play and gives not in to Paris' lust. In other words, the goddess of love threatens to make Helen be perceived as the real cause of evil and not visible to Trojan eyes through the beguiling illusion of isolated erotic desire. The cowed Helen complies; and, back in her chamber, Paris, awaiting, again seduces her by using the argument that war

---

[5] Voegelin 97. The term has erotic reference also when Hera suggests to Zeus that no one should see them engaged in love making as the god draws the clouds over them [*Il*. XIV, 336]. There occur in the narrative other uses for the term meaning *shameful blame*. In Phoenix's *paraenesis* to Achilles during the embassy his aged tutor reminds the hero that there was no *shameful blame* for his anger until now that his fellow warriors beseech him offering gifts from Agamemnon to control his wrath [*Il*. IX, 523], which he eventually subdues. Odysseus himself offers consolation to Agamemnon reminding him that there is no *shame* in offering gifts to make up for past ills even for a king, at the time Achilles gets ready to rejoin the fighting [*Il*. XIX, 182]. Lastly, Hermes addresses Priam, after accompanying him through enemy lines on his way to ransom Hector's corpse; the god leaves the King of Troy to visit Achilles on his own, because it would be *shameful* for a god to be entertained by a mortal [*Il*. XXIV, 463]. The term is used when public blame could be incurred through improper behavior by gods and mortals. In the case of Helen there is no effective avoidance since she gives in to *shame* when threatened by Aphrodite.

25

has intensified love in his soul and his desire for her is stronger than ever [*Il*. III, 442-446]. In this dream play Paris cheats death through the intensity of erotic desire, personified in the blind Trojan worship of Aphrodite.[6]

Helen appears again in the *Iliad* in Book VI. Hector visits Paris in his chamber and rebukes him for failing to join the others in battle [*Il*. VI, 343-368]. Paris justifies his indolence by insisting that he has been grieving in sorrow over the constant slaughter and agrees to rejoin the fighting. In this scene Helen beckons Hector to sit by her and rest from all the warfare taking place "for the sake of shamed me and the blind passion of Alexandros" [*Il*. VI, 356]. She adds that Zeus has set on them a hard fate so that they may be made the subject of song henceforth [*Il*. VI, 358]. Helen and Hector both know the great dimension of the important roles they must play in the heroic epic. The worship of Aphrodite exacts its toll.

In Book VII Helen does not appear herself, but figures prominently in the discussion. There is an assembly held within the walls of Troy. Antenor, an aged counsellor, the counterpart of Nestor in the Greek camp, proposes that Helen of Argos be given back to the grieving Achaeans, along with all her possessions, since "we now fight with our sworn words turned into lies [*Il*. VII, 351-352]. [7] Antenor is probably referring to the duel between Menelaus and Paris which was intended as a trial by combat to settle the issue and end the war, but which Aphrodite's dream play interrupts by whisking away Paris into Helen's bedchamber. The words of the aged counselor may be taken to refer indirectly also to the original abduction of Helen by Paris while a guest at her husband's royal court.[8] In either case Paris feels spurned by the argument raised over the

---

[6] Voegelin considers narration of the combat between Menelaus and Paris in the Third Book of the *Iliad* "interpenetration of tragedy and comedy." Voegelin 93. The trial by combat could provide resolution for a struggle which is not meant to end. Voegelin 94. In the confusion caused by the sudden absence of Paris from the battlefield due to Aphrodite's dream play, a Trojan takes a potshot at Menelaus and the trial by combat truce is broken. [*Il*. IV, 85-222]

[7] νῦν δ᾽ ὅρκια πιστὰ/ ψευσάμενοι μαχόμεσθα.

[8] Later in the narrative, there is a Trojan warrior, Hector's scout, who epitomizes the Trojan vain use of words. When decapitated by Odysseus, Dolon's head rolls downhill still speaking though separated from his body, a symbolic representation of the Trojan persistent, useless

object of his erotic desire and refuses outright to give back "his wife," *gunaika*, although he is willing to give back the treasures he took, adding additional compensation from his store [*Il.* VII, 362-364]. Fairness of the crucial proposal, with the important partial refusal, is not considered by members of the assembly as a parliamentary issue whose justice can be questioned, discussed, and voted on; rather, Priam stands and asks that Paris' uncontested reply be delivered by a herald to the Greeks, adding a request that a truce be granted for burial of the dead on both sides. "Paris' decision, for whose sake the war started," is to be delivered to the fighting Danaans [*Il.* VII, 374]. [9] The herald delivers the message, quoting verbatim Priam's epithet for Paris [*Il.* VII, 388]. The messenger adds a parenthetical embellishment all his own; referring to Paris, he wishes that he should have died before [*Il.* VII, 390]. [10] The importance of the announcement is given legal force since the herald uses in place of Helen's name an epithet, "wedded wife of glorious Menelaus" [Il. VII, 392]. [11] The message includes also what is probably a truism; Paris refuses to give back Helen, but the Trojans wish he would. [*Il.* VII, 393]

At the Greek Assembly Diomedes is quick to speak in order to refuse the proposed partial compensation, probably because it excludes the return of Helen, using prominently the lady's name, conspicuously absent from Paris' and the herald's speech [*Il.* VII, 401]. Homer may be pointing out the Trojan guilt implicitly in such obfuscated reference to Helen. Agamemnon echoes the declaimed refusal to accept partial compensation but grants the truce for burial of the slain on both sides.

Guilt for the Achaean-Trojan struggle is attributed to Helen several times. Hera, cohort of Zeus, tells Athena to inspire courage on the Argives when they

---

verbiage intended to distract the listener and not resolve the issue of Helen's unlawful abduction. [*Il.* X, 457]

[9] μῦθον Ἀλεξάνδροιο, τοῦ εἵνεκα νεῖκος ὄρωρε. [*Il.* VII, 374; *Il.* VII, 388.]

[10] ὡς πρὶν ὤφελλ' ἀπολέσθαι [*Il.* VII, 390]. Helen expresses the very same wish applied to herself in her elegiac lament over Hector. [*Il.* XXIV, 764]

[11] κουριδίην δ' ἄλοχον Μενελάου κυδαλίμοιο.

27

are yearning to sail back to Greece; they should stay, fight, and not leave to Priam and to the Trojans Helen of Argos, because "many have lost their lives in nine years of fighting for her sake" [*Il*. II, 161-162]. Athena goes to Odysseus and delivers the message quoting the same line of verse [*Il*. II, 176-177]. Odysseus brandishes Agamemnon's scepter as sign of authority when addressing noblemen and strikes with it common warriors while reprimanding them for fleeing, despite Hera's request to stay and fight. Helen's rapture is also used as a point of leverage to spurn warriors to continue on with the struggle. Nestor, the aged counselor, stirs the troops by urging the men to fight for victory "so each one of the Greeks may lay in bed with the wife of a Trojan to avenge the grief of Helen" [*Il*. II, 354-356]. Adequate retribution appears to be the release of sexual passion. When exposing the catalogue of ships included in the Greek naval force, Homer uses the same line of verse indicating revenge through sexual release to justify Menelaus' resentment in a war for the loss of Helen [*Il*. II, 590]. Abduction engenders a presumed need for further rapture as the wheel of fortune spins uncontrollably.

When appearing in the narrative, the character of Helen is self-conscious about her compromising position. She refers to herself as "shameless me, to whom Agamemnon was once a brother-in-law," when pointing out to Priam the fighting heroes before the dream play in the Third Book [*Il*. III, 180]. [12] Helen's shame is most obvious in the scene where, addressing Hector as her "brother-in-law," she calls herself a despised mischievous bitch [*Il*. VI, 344]. [13] This self-deprecating taunt appears an indirect effort to appeal to Hector's carnal desire. Noble Hector reacts by dashing off rapidly to seek the company of his wife [*Il*. VI, 365-366]. Compared to Helen and Paris, Hector's character is beyond reproach.

---

[12] δαὴρ αὖτ᾽ ἐμὸς ἔσκε κυνώπιδος. In his *Homeric Lexicon* Richard J. Cunliffe considers κυνώπιδος as an instance of self-deprecation. Cunliffe 242. Herbert W. Smyth in his *Greek Grammar* mentions that we find in Homer iterative imperfect tenses "denoting a customary or repeated action." Smyth 146. The adjective κυνώπιδος, dog eyed, used with the verb "to be" means that Helen remains "lost to all sense of decency." Cunliffe 242.

[13] δᾶερ ἐμεῖο κυνὸς κακομηχάνου ὀκρυοέσσης. The term κακομηχάνου means "contriving evil." Cunliffe 207. Helen is distasteful to the point of "causing shuddering," ὀκρυοέσσης. Cunliffe 289.

Perhaps the most damning reference to Helen as cause for the conflict is the bitter, open lamentation of Achilles when he mourns the death of Patroclus in Book XIX. The hero exclaims that neither the death of his aged father, Peleus, nor that of his son Neoptolemus, not even his own death, in a foreign land "over horrible Helen," εἵνεκα ῥιγεδανῆς Ἑλένης, would grieve him more than Patroclus' death [*Il*. XIX, 325]. In the *Iliad* Achilles is destined to suffer punishment for his resentful, and costly, withdrawal from battle by enduring the loss of Patroclus. Although Achilles kills Hector, slayer of his beloved comrade, the abduction of Helen always lurks in the background as original pivoting point to absorb the real, ethical cause for the devastation of war. The famous beauty is conscious of her compromising position, yet Helen's most tender psychological lament comes at the end of the *Iliad*. After Andromache and then Hecuba mourn the dead Hector [*Il*. XXIV, 725-745; *Il*. XXIV, 747-759], Helen presents a heart-felt litany over the slain noble hero. [*Il*. XXIV, 762-775]

The Argives must establish complete differentiation from the Trojans. Eventually, Agamemnon recognizes the need to quell Achilles' anger over the absence of Briseis. Consequently, the Greek chieftain sends an embassy to Achilles. Odysseus Ajax, and Phoenix, Achilles' old tutor, try to persuade the sulking hero to accept restitution and return to battle. Restoration to order, as outlined by the fable of the *Litai*, Spirits of Prayer, in Phoenix's *parainesis* to Achilles during the embassy to his tent in Book IX, offering restitution, requires human acceptance of great physical loss with humility and resignation:

> For there are also the spirits of Prayer, the daughters of great Zeus
> and they are lame of their feet, and wrinkled, and cast their eyes sidelong,
> who toil on their way left far behind by the spirit of Ruin:
> but she, Ruin, is strong and sound on her feet, and therefore
> far outruns all Prayers, and wins into every country
> to force men astray; and the Prayers follow as healers after her.
> If a man venerates these daughters of Zeus as they draw near,
> such a man they bring great advantage, and hear his entreaty;
> but if a man shall deny them, and stubbornly with a harsh word
> refuse, they go to Zeus, son of Kronos, in supplication

that Ruin may overtake this man, that he be hurt, and punished.
[*Il*. IX, 502-512] [14]

καὶ γάρ τε λιταί εἰσι Διὸς κοῦραι μεγάλοιο
χωλαί τε ῥυσαί τε παραβλῶπές τ᾽ ὀφθαλμώ,
αἵ ῥά τε καὶ μετόπισθ᾽ ἄτης ἀλέγουσι κιοῦσαι.
ἡ δ᾽ ἄτη σθεναρή τε καὶ ἀρτίπος, οὕνεκα πάσας
πολλὸν ὑπεκπροθέει, φθάνει δέ τε πᾶσαν ἐπ᾽ αῖαν
βλάπτουσ᾽ ἀνθρώπους· αἱ δ᾽ ἐξακέονται ὀπίσσω.
ὃς μέν τ᾽ αἰδέσεται κούρας Διὸς ἆσσον ἰούσας,
τὸν δὲ μέγ᾽ ὤνησαν καί τ᾽ ἔκλυον εὐχομένοιο·
ὃς δέ κ᾽ ἀνήνηται καί τε στερεῶς ἀποείπῃ,
λίσσονται δ᾽ ἄρα ταί γε Δία Κρονίωνα κιοῦσαι
τῷ ἄτην ἅμ᾽ ἕπεσθαι, ἵνα βλαφθεὶς ἀποτίσῃ.
[*Il*. IX, 502-512]

Achilles refuses the restitution offered by the embassy, displaying excessive, unmitigated bitterness. Eric Voegelin explains the fable of the *Litai* [502], Prayers, daughters of Zeus [508], as the dialectics of guilt, *Ate* [504, 505, 512]:

> The Homeric *Ate* means the folly of the heart, the blindness
> of passion, that makes a man fall into guilt; and it also means
> the sinful act, the transgression of the law. And the Homeric *Litai*
> correspondingly means the repentance of the heart, as well as the acts
> (prayers and sacrifices to god, prayers and offers of recompense to men)
> in which repentance expresses itself. *Litai* are the daughters of Zeus
> in so far as they express the active willingness to rise from the fall
> into disorder, to heal the guilt… Hence if a man repels another
> man's manifest willingness to repair the broken order, he himself
> falls into the guilt of perpetrating disorder; the disorder is now his
> *Ate* for which he will have to make full atonement. [15]

---

[14] Lattimore 211. The *Litai* are personified "prayers of repentance, for forgiveness, addressed by an offender to the person injured." Cunliffe 251. The "perversion or deception of the mind leading to evil doing," is also personified as *Ate*, or "Ruin" in Lattimore's translation. Cunliffe 59.

[15] Voegelin 87-88.

Submission to the peaceful entreaty of a suppliant cannot be refused. Suppliants cannot be cast out, nor ignored. Achilles must heed the claim of Agamemnon's embassy. The ambassadors include Odysseus. We know that in the *Odyssey* the hero endures shipwreck and the loss of his men until he initiates return to his own land at Alcinous' court. Princess Nausicaa reminds him of a chaste young Penelope. [16] His return to Ithaka involves adopting a veritable figure of the *Litai*; Odysseus becomes a beggar who enlists the aid of a swine herd. Surreptitiously he infiltrates the royal court to punish Penelope's suitors and reclaim his throne.[17] Suppliants are protected by Zeus,

A legitimate gripe must be recognized, but to go overboard in the reaction is forbidden. The question arises as to exactly how far the injury suffered is a cause justifiable enough for a balanced reaction. The quest for equitable restitution links the legend of the Trojan war to the *Iliad* as an epic. On the one hand, the Greeks sail away from their homeland and lay siege to Troy seeking compensation for the abduction of Helen and her treasures; on the other hand, Achilles withdraws from the fighting due to the sequestering of Briseis by Agamemnon. Suing for redress is the start of implementing justice; the path for seeking appropriate restitution paves the way for civilized behavior. The nature of the quarrel between Menelaus and Paris over Helen occasions greater loss than Odysseus' shipwreck; although Penelope's suitors echo the sentiment that adulterous lust is an analogy for breach of political loyalty. The Trojan war causes extreme suffering, exemplified by the death of Hector and countless others. The issue of Achilles sulking away due to Agamemnon's taking of Briseis results in the death of Patroclus and the excessive number of Achaeans lost due to the abstinence from battle of the greatest warrior. [18] Achilles' wrath is justifiable, yet

---

[16] The Phaeacian royal family exudes symbolic perfection. The mother of the princess, and the wife of Alcinous, the Phaeacian king, is named *Areté*, meaning "virtue." Jaeger mentions that the name choice for the Phaeacian queen reveals the educative influence of women upon a stern and warlike masculine society. Jaeger 23.

[17] Lattimore's *Odyssey*, Book 22.

[18] Jaeger observes how the fate of Patroclus and the Greek army depend on Achilles' anger evoked by the "mighty religious conception of *Ate*, the madness of doom: like an ominous

compensable. Although Achilles does not accept immediate compensation from Agamemnon's embassy, nevertheless he keeps Phoenix, his old tutor, by his tent.[19] Indirectly, Achilles shows that the fable of the *Litai* has persuaded him enough to contemplate submission to his fate; as warrior he must fulfil the prophecy of dying young with glory.

Both Hector and Patroclus represent the sacrifice suffered for the sake of appeasing a loss. How much should a man pay to attain the object of his desire is in question. Gans accounts for resentment as the bitterness expressed for "induced excess of appetite for the central object." [20] Paris, the archer who fights from a distance, avoids responsibility for his actions. Achilles, the swordsman, must face his guilt and accept that the product of struggle can be reconciliation. Paris remains in the legend. Achilles takes up his symbolic place, along with Hector, at the crux of Western civilization.

Resolution of a conflict comes through sacrifice. Retribution comes after the slaying of Patroclus, Achilles' *alter ego*, and at the death of Hector, Priam's son, and Paris' older brother. The legend and the epic must be aligned in our understanding. Although Briseis is not as prominent in the narration, unlike the several discussed scenes in which Helen's persona enters the narration, Briseis nevertheless is a prominent place holder in the *Iliad* accounting for Achilles' wrath and his subsequent withdrawal from the fighting. Her elegiac mourning over Patroclus parallels Helen's mourning over Hector. Patroclus is to Briseis what Hector is to Helen, a kind benefactor. Thus, we may include Patroclus in the love triangle between Agamemnon and Achilles over Briseis, just as Hector enters

---

phantom, it rises half-seen behind the other compelling allegory of the *Litai*, the prayers and the callous heart of man." Jaeger 27.

[19] Phoenix's identity is to stand as Achilles' teacher. Jaeger 8. The old counselor advises the young warrior "to accept Agamemnon's gift as an atonement." Jaeger 10. The historian explains further that "the courage of a Homeric nobleman is superior to a berserk contempt of death in this: that he subordinates his physical self to the demons of a higher aim, the beautiful. And so the man who gives up his life to win the beautiful, will find that his natural instinct for self-assertion finds its highest expression in self-sacrifice." Jaeger 13. We observe that in Classical Antiquity to combine self-assertion with self-sacrifice is a unique pre-Christian phenomenon.

[20] Signs 143.

as paradigm of loss to be suffered and endured in the struggle of Menelaus and Paris over Helen. Homer possibly intended, in his architectonic conception of the *Iliad*, that the love triangle should be expanded and better viewed as a quadrangular relationship which includes the theme of loss and deadly suffering for the appeasement of desire. [21]

The cornerstone speech outlining the quadrangular conception occurs when Briseis mourns the dead Patroclus as the one who would have made her Achilles' legally wedded wife:

> Patroclus, far most pleasing to my heart in its sorrows,
> I left you here alive when I went away from the shelter,
> but now I come back, lord of the people, to find you have fallen.
> So evil in my life takes over from evil forever.
> The husband on whom my father and honored mother bestowed me
> I saw before my city lying torn with the sharp bronze,
> and my three brothers, whom a single mother bore with me
> and who were close to me, all went on one day to destruction.
> And yet you would not let me, when swift Achilles had cut down
> my husband, and sacked the city of godlike Mynes, you would not
> let me sorrow, but said you would take me back in the ships
> to Pythia and formalize my marriage among the Myrmidons.
> Therefore, I weep your death without ceasing. You were kind always. [22]
> [*Il.* XIX, 287-300]

> Πάτροκλέ μοι δειλῇ πλεῖστον κεχαρισμένε θυμῷ
> ζωόν μέν σε ἔλειπον ἐγὼ κλισίηθεν ἰοῦσα,
> νῦν δέ σε τεθνηῶτα κιχάνομαι ὄρχαμε λαῶν
> ἄψ ἀνιοῦσ᾿· ὥς μοι δέχεται κακὸν ἐκ κακοῦ αἰεί.
> ἄνδρα μὲν ᾧ ἔδοσάν με πατὴρ καὶ πότνια μήτηρ
> εἶδον πρὸ πτόλιος δεδαϊγμένον ὀξέϊ χαλκῷ,
> τρεῖς τε κασιγνήτους, τούς μοι μία γείνατο μήτηρ,
> κηδείους, οἳ πάντες ὀλέθριον ἦμαρ ἐπέσπον.
> οὐδὲ μὲν οὐδέ μ᾿ ἔασκες, ὅτ᾿ ἄνδρ᾿ ἐμὸν ὠκὺς Ἀχιλλεὺς
> ἔκτεινεν, πέρσεν δὲ πόλιν θείοιο Μύνητος,

---

[21] René Girard in his *Violence and the Sacred* explains the *mimetic double bind* as the persistent rivalry among subjects with a similar object of desire. In the typical love triangle, the subjects with a similar goal but conflicting need are rivals in the quest for appropriation of the chosen object. Girard 178-182, 186, 188.

[22] Lattimore's *Iliad* 399-400.

κλαίειν, ἀλλά μ᾽ ἔφασκες Ἀχιλλῆος θείοιο
κουριδίην ἄλοχον θήσειν, ἄξειν τ᾽ ἐνὶ νηυσὶν
ἐς Φθίην, δαίσειν δὲ γάμον μετὰ Μυρμιδόνεσσι.
τώ σ᾽ ἄμοτον κλαίω τεθνηότα μείλιχον αἰεί.
[*Il*. XIX, 287-300]

Patroclus is the kind benefactor who leads Briseis to hope for a union in marriage with Achilles. Widowhood makes her a more veridic figurehead for wedded wife than Helen. The character of Patroclus shows a kindness which we can identify as exceptional and is ratified by his prominence in the plot. In Book XI Paris shoots and strikes Machaon, the healer, son of the great physician Asclepius [*Il*. XI, 506]. Achilles observes from a distance a fallen warrior in the battlefield who seems to him to be Machaon and sends Patroclus to verify that the healer has fallen [*Il*. XI, 610-612]. Nestor had already carried Machaon off in great sorrow, for a healer is the one man who is worth many men due to his unique ability to remove arrows and apply medicine [*Il*. XI, 514-515]. Achilles' closest friend verifies the fact that Machaon has been struck [*Il*. XI, 650]. Patroclus himself heals Eurypilus [*Il*. XII, 2; *Il*. XV, 394]. We should notice that the original intent for Patroclus to leave Achilles' camp and enter the melee is concern over Machaon's absence due to injury in battle. But offering aid to the fallen is not enough for noble Patroclus. The hero, weeping bitterly, returns to Achilles. The great warrior scolds his friend for crying like a little girl [*Il*. XVI, 7]. At this point in the narrative Patroclus ushers forth a bitter reprimand to Achilles. His parents cannot be Thetis and Peleus, but rather the grey sea and the jagged cliffs must have engendered him [*Il*. XVI, 33-35]. Then noble Patroclus asks Achilles for his armor and weapons so he can return to battle in his stead to frighten the Trojans into thinking that he is him [*Il*. XVI, 41].[23] Events leading to Achilles' return to battle originate in the wounding of Machaon, the healer. Patroclus assumes the role of savior.

---

[23] Nestor had told Patroclus to ask for Achilles' armor as disguise and to join the fighting leading a troop of Myrmidons to battle in his place since the young warrior was held back by a prophecy from his mother [*Il*. XI 793-799]. The aged counselor considers that persuasion by Patroclus, Achilles' older friend, is vital. [*Il*. XI, 786, 790, 792]

Achilles considers Patroclus' request seriously. His response outlines in detail the nature of his stance and sets the tone for what follows. In his answer to Patroclus, Achilles expresses resentment because Agamemnon had placed him in the role of a "dishonored outcast" [*Il*. XVI, 59]. The hero had referred to himself by using the same self-deprecating epithet in his answer to Ajax's plea, begging him to return to the battlefield, during the embassy from Agamemnon in the Ninth Book [*Il*. IX, 648]. Repetition of the exact same term, ἀτίμητον μετανάστην,[24] reveals how Achilles feels contempt over the role of suppliant, customarily adopted by someone seeking proper redress from a wrong suffered. As he sends Patroclus, wearing his own armor, to take his place in the fighting, Achilles tells his dear friend to come back to his encampment as soon as the Trojans are close to the ships [*Il*. XVI, 95]. His wish is that, at the moment preceding the instance of ultimate disastrous defeat for the Argives, he can claim to be the Greeks' savior along with Patroclus [*Il*. XVI, 100]. Here pride transcends all rationale. Achilles must accept that he cannot reach the glory he craves in life; such is the nature of his prophetic death in youth. He must experience the loss of his *alter ego*, κεφαλῇ [*Il*. XVIII. 82], before he can accept the duty of his calling: to perform as the greatest warrior, regardless of reward. [25] In order to attain eternal glory, he must not chide his share in mortality. His friend cannot avoid demise. Patroclus' death at the hands of Hector is the loss Achilles must suffer to realize who he is. Fame in unison with his best friend is a reward in the afterlife. Dying young with glory includes foregoing the pleasures of old age among mortals. Their heroic triumph is not terrestrial.

Hector is another victim who must perish in the story to outline the sacrificial nature of mimetic desire in Homer. The noble warrior is a character

---

[24] Ἀτρεΐδης ὡς εἴ τιν' ἀτίμητον μετανάστην [*Il*. IX, 648; *Il*. XVI, 59]. The two lines where the unique epithet in question appears are identical in the Homeric text. The term, μετανάστην, is defined as "one who has changed his home, an exile, an outcast." Cunliffe 266.

[25] Voegelin 92.

who places family matters first. [26] This role prevails when Hector mourns fallen heroes in Book Thirteen [*Il.* XIII, 769-773]. He reprimands Paris as a womanizer but refrains from any further chastise. Paris retorts that Hector should not blame the blameless, ἀναίτιον αἰτιάασθαι [*Il.* XIII, 775]. Although the great warrior acquiesces, yet the subsequent death of Hector constitutes a tragedy. Mirroring Briseis' lament over Patroclus, Helen starts her litany over Hector's corpse by considering herself the wife of Paris:

> Hector, of all my lord's brothers dearest by far to my spirit:
> my husband is Alexandros, like an immortal, who brought me
> here to Troy; and I should have died before I came with him;
> and here now is the twentieth year upon me since I came
> from the place where I was, forsaking the land of my fathers. In this time
> I have never heard a harsh saying from you, nor an insult.
> No, but when another, one of my lord's brothers or sisters, a fair-robed
> wife or some brother, would say a harsh word to me in the palace,
> or my lord's mother – but his father was gentle always, a father
> indeed – then you would speak and put them off and restrain them.
> I mourn for you in sorrow of heart and mourn myself also
> and my ill luck. There was no other in all the wide Troad
> who was kind to me and my friend; all others shrank when they saw me.
> [*Il.* XXIV, 762-775] [27]

> Ἕκτορ ἐμῷ θυμῷ δαέρων πολὺ φίλτατε πάντων,
> ἦ μέν μοι πόσις ἐστὶν Ἀλέξανδρος θεοειδής,
> ὅς μ᾽ ἄγαγε Τροίηνδ᾽: ὡς πρὶν ὤφελλον ὀλέσθαι.
> ἤδη γὰρ νῦν μοι τόδε εἰκοστὸν ἔτος ἐστὶν
> ἐξ οὗ κεῖθεν ἔβην καὶ ἐμῆς ἀπελήλυθα πάτρης:
> ἀλλ᾽ οὔ πω σεῦ ἄκουσα κακὸν ἔπος οὐδ᾽ ἀσύφηλον:
> ἀλλ᾽ εἴ τίς με καὶ ἄλλος ἐνὶ μεγάροισιν ἐνίπτοι
> δαέρων ἢ γαλόων ἢ εἰνατέρων εὐπέπλων,
> ἢ ἑκυρή, ἑκυρὸς δὲ πατὴρ ὣς ἤπιος αἰεί,
> ἀλλὰ σὺ τὸν ἐπέεσσι παραιφάμενος κατέρυκες
> σῇ τ᾽ ἀγανοφροσύνῃ καὶ σοῖς ἀγανοῖς ἐπέεσσι.
> τὼ σέ θ᾽ ἅμα κλαίω καὶ ἔμ᾽ ἄμμορον ἀχνυμένη κῆρ:

---

[26] Considering Hector as defender of the hearth, Jaeger quotes the hero's advice to men from Troy: "One omen is the best: fight for your home" [*Il.* XII, 243]. Jaeger 45.
[27] Lattimore 495.

οὐ γάρ τίς μοι ἔτ᾽ ἄλλος ἐνὶ Τροίη εὐρείη
ἤπιος οὐδὲ φίλος, πάντες δέ με πεφρίκασιν.
[Il. XXIV, 762-775]

To justify her legitimate claim to mourn for Hector, Helen calls Paris her husband, πόσις [Il. XXIV, 763].[28] In her view of life at Troy Paris' siblings become her brothers-in-law [Il. XXIV, 769]. Among them Hector is her kindest benefactor [Il. XXIV, 762]. Helen feels guilt and wishes she had died before coming to Troy [Il. XXIV, 764]. Hector's protection of Helen extends to the treatment he expected her to receive as an equal, among his other relatives. Alliance to the family legitimizes her presence as a companion to Paris at Troy. Helen alludes to a difference between Hecuba and Priam, this difference is most obvious when Hecuba discloses her bitterness over Hector's death. [29] Although Hector reprimands Paris, he welcomes Helen to the family. In her mournful elegiac speech, she expresses intense gratitude at Hector's unique kindness.

Patroclus and Hector fall as victims sacrificed to the mimetic double bind conflict among embittered rivals over Briseis and Helen. The two warriors closest to the heroes Achilles and Paris, represent the ransom exacted by desire in the *Iliad*. The relationships involved in linking Briseis to Patroclus and Helen to Hector are balanced by the paradox of opposites. Briseis, born in the Troad, is protected by the Greek Patroclus; while Helen, a Greek, is protected by Hector, a Trojan. This opposing binary structure supports the direct equivalence we ascribe to the importance of two leading women characters, absent from their clan, who subsist among antagonistic factions in the *Iliad*. The dispossession of Achilles' Briseis, identified as the warrior's prize [Il. I, 185], mirrors in microcosm the legendary abduction of Helen by Paris while a guest at Menelaus' court, presumed

---

[28] She refers to Menelaus as her former husband, πρότερον πόσις [Il. III 167, 429]. Briseis uses the term ἀνήρ to refer to her husband [Il. XIX 291, 295]. This last term includes the sense of being a mature man, not a youth, along with the meaning of husband.

[29] Hecuba tells Priam she wants to gnaw at Achilles' liver when the Trojan King is drawing plans to ransom Hector's corpse. [Il. XXIV, 213]

cause for the Trojan war. To maintain the role of victors in the struggle, Argives must differentiate themselves from their Trojan foes. [30] Upon seizing Briseis, Agamemnon is indirectly aligning himself with the enemy, hence justifying Achilles' struggle for differentiation through rightful indignation – even though Briseis herself was obtained by the great warrior upon his slaying of her husband, conforming to the typical acquisition of spoils in fair battle during persistent warfare in a primitive tribal society. The deaths of Hector and Patroclus fulfill the ransom exacted by the desires of Paris for Helen and Agamemnon for Brises, respectively.

The expurgation of Achilles' anger, *Menis*, is the central theme of the *Iliad*, as Homer tells us by choosing the term as the first word of the poem. [31] The price demanded by fate comes to fruition in Book XVIII of the *Iliad*. Voegelin believes that: "the drama of the *cholos* hinges on the death of Patroclus; with the death of his friend the obsession of Achilles falls apart, and the reality of life and order is restored." [32] Achilles' return to the fighting involves recognition that Agamemnon angered him, ἐχόλωσεν [*Il*. XVIII, 111], and gave rise to his indignation. Now Hector must die to avenge the slain Patroclus; Achilles must fight and face his own death. The great warrior deeply laments the toll exacted by fate due to his refusal to join the melee: loss of his dear friend, φίλης κεφαλῆς [*Il*. XVIII, 114]. [33] Voegelin considers the dialogue between Achilles and his mother, the

---

[30] Girard explains that impartiality "sees the roots of justice in differences among men and the demise of justice in the elimination of these differences." Girard 51.

[31] Although *menis* in the first line of the *Iliad* characterizes the wrath of Achilles, the term also applies to the general temper of a people in Hesiod, *Scutum Herculis*, 21. Liddell and Scott 1128. *Menis* is translated often as ire. Cunliffe 269. In the *Agamemnon* of Aeschylus *menis* refers to the anger of injured parents. *Ag*. 155. Liddell and Scott 1128. *Cholos*, on the other hand, refers to bitter gall, the denotation is more deeply psychological. Lidell and Scott 1997. This last term expresses giving in to the effects of anger. Cunliffe 420.

[32] Voegelin 92. The critic explains *cholos* as being analogous to the Latin *inimicitia*, and similar to a Medieval feud. Voegelin 89. Curiously enough, the first mention of *cholos* in the *Iliad* is in reference to Apollos's anger against Agammemnon for dishonoring Chrises. [*Il*. I, 9]

[33] The remarkable closeness of the relationship between Achilles and Patroclus may not be as accessible to a modern reader as it was to Homer's audience. A.W.H. Adkins explains the meaning of the adjective modifying κεφαλῆς, literally, head. "It is not surprising that Homeric man should use some word to demarcate the persons and things on which his existence depends

nymph-goddess Thetis, where Homer expounds on the reasons for the hero's return to the fighting [*Il*. XVIII, 78-126], central to the narrative; the critic calls this passage "the psychological masterpiece of the *Iliad*." [34] After this climactic peak, Achilles returns to the fighting because Patroclus has died, and the great hero wants to join his friend in the afterlife. He wants Hector dead and is ready to die himself soon, fulfilling the prophecy that rules his destiny as warrior, if Hector pays for Patroclus' death. Patroclus' shade tells Achilles that death is near, and both their ashes must be placed in the same urn [*Il*. XXIV, 91-92]. There is a new test for the measure of Achilles' indignation. His vengeance must reach a true measure, commensurate with the suffering caused by Patroclus' death. Revenge over Patroclus' death gains significance when balanced by the loss of Hector, his slayer.

The gods are sympathetic. Achilles slays Hector, yet the Trojan leader's body cannot be disfigured while being dragged around mercilessly. The gods prevent decomposition until Priam arrives to ransom the princely corpse of his warrior son. The moment of requital harkens back to the early scene when Briseis is taken. At that moment Achilles is restrained by Athena from drawing his sword against Agamemnon [*Il*. I, 194-195]. Gans reminds us that: "The wrath of Achilles is no mere model of the Greeks' wrath against Troy for the theft of Helen. It is an expression of resentment, not of heroic rivalry, and its effect is to transform the heroic narrative into literature." [35] Achilles' refusal to battle Agammemnon is explained: "The necessary choice of 'wrath' over violence is an archetypal genesis of resentment." [36] Thus, the *Iliad* starts with the proposal to dwell on the hero's anger. Likewise, at the end of the epic, Achilles must subdue his wrath and refrain

---

and distinguish them from persons and things in general. The word is *philos*, conventionally rendered 'own' or 'dear;" in fact untranslatable, for we are not acutely aware of possessing a limited stock of persons and things upon whom our very existence depends." Adkins 16. Cyrus H. Gordon considers David's mourning over his beloved friend Jonathan in the *Old Testament* [2 Sam. 22] like Achilles' sentiment for Patroclus. Gordon 71, 272.

[34] Voegelin 92.

[35] Culture 244.

[36] Culture 246.

from striking down Priam when the Trojan King insists on displaying excessive eagerness to bury Hector's corpse while he is a guest at the hero's tent [*Il.* XXIV, 560-570]. The king must bargain patiently for his son's body, so it can be taken back to Troy for public burial. Achilles grants a grieving father the body of his slain son, which is the hero's battle trophy. [37] Each party has a notion different from what the other is getting, but the transaction is effectuated.

Gans mentions that the *Iliad*, today a work of fiction, was earlier recited at religious festivals as the equivalent of "oral scripture." [38] Voegelin compares composition of the Biblical narrative and the Homeric epic by explaining that the "traditional-historical" method in the former and the "oral composition" of the latter make the architecture and meaning of the works compatible.[39] In the Biblical narrative religious practice includes instructions on how to worship and how to live.[40] The Homeric epic could also express beliefs and customs generally seen in the context of religion and law. Burial of Hector's corpse brings together the will of the gods and the customary respect owed the dead. [41]

---

[37] Otto Gierke in his *Associations and Law* explains that in Classical Antiquity there is an "imperative postulate contained in the idea of law which obliges the ruler to utilize the power not for himself but for the community." Gierke 83.

[38] Chronicle 771.

[39] Voegelin 69.

[40] Religious worship and the rule of law is combined in the *Ten Commandments*. Exodus 20: 2-17. Bullinger 100-101.

[41] Early in Book XXIV, Apollo in Olympus pleads that the burial of Hector be allowed using the term *nekus*, corpse [35]. In the same Book, Zeus tells Thetis to communicate to Achilles the need for the corpse to be ransomed [108]. Achilles also refers to Hector's corpse as *nekus* [423] several times. Addressing Priam, Hermes pairs *uios* [422], son, in apposition to the term to indicate that the gods care for Hector, although he is a corpse. Achilles orders a tunic to wrap the body for Priam to take back to Troy [581]. The warrior advises Priam to be cautious on his way back, lest someone sees him and tells Agamemnon, for then a delay over the exchange might occur [655]. Homer describes the corpse being taken back on a mule after Hermes' departure [697]. During the actual exchange, Achilles urges Priam to be patient while the bargain goes on referring to the corpse as "Hector" [561]. He even asks Priam how many days are needed for "Hector's" burial [657]. Priam asks for a twelve-day truce for the ceremony [660], which is granted by Achilles. In his dialogue with Hermes, before meeting Achilles, Priam uses the term *pais*, child, to ask whether his son has been exposed to scavenging dogs [408]. Later Hermes reassures Priam that his ransomed son, υἱὸν ἐλύσαο, can be taken back to Troy safely [685]. Priam describes the final burial for "him," *autov* [665, 667], Hector. We notice that Achilles refrains twice from referring to Hector's corpse as a cadaver during final

Priam and Achilles share an intimate dialogue in which the warrior observes that the condition of mortals is intrinsically much different from the god's stance as immortals:

> Such is the way the gods spun life for unfortunate mortals,
> That we live in unhappiness, but the gods themselves have no sorrows.[42]
> [*Il.* XXIV, 525-525]

> γὰρ ἐπεκλώσαντο θεοὶ δειλοῖσι βροτοῖσι
> ζώειν ἀχνυμένοις: αὐτοὶ δέ τ᾽ ἀκηδέες εἰσί.
> [*Il.* XXIV, 525-526]

The gods spin the web for human sorrow, while they themselves remain aloof from the bereavement endured by the loss of a loved one. Priam has lost Hector; Achilles has lost Patroclus. Destiny brings the king and the warrior together. The mutual intimacy evoked in the confrontation between Priam and Achilles is crucial. Woeful anguish unites them in lamentation. Priam wants his son's corpse to take to Troy for public mourning. Achilles must turn over Hector's corpse to subdue his anger and put away his thirst for vengeance. Although each party in the agreement gets something different from what the other bargains for, Hector's corpse is ransomed. [43] The Trojan warrior's corpse represents the additional third

---

negotiation over the body of Hector in lines 561 and 657, displaying tact in order to consummate the transaction.

[42] Lattimore 489. The term ἀκηδέες [526] is defined by Cunliffe as having the sense "vexed by no cares." Cunliffe 16. The adjective is formed from the negation of the noun κῆδος, meaning "trouble or grief occasioned by a family loss." Cunliffe 226.

[43] Homer covers all aspects of the transaction involved in the contractual negotiations leading to the ransom of Hector's corpse. 1) Zeus tells Hera to summon Thetis [*Il.* XXIV, 65-76]; 2) Hera draws Thetis over to Zeus [XXIV, 88]. 3) Zeus talks to Thetis: Achilles must give back Hector's body for burial [XXIV, 104-119]. 4) Thetis persuades her son to give back Hector's body [XXIV, 128-137]. 5) The hero assents in the same Book [139-140]. 6) Zeus sends for Iris; Achilles must spare a suppliant, [144-158]. 7) Iris reassures Priam [171-187]. 8) Priam tells Hecuba his intention to approach Achilles [194-199]. 9) His wife warns Priam about the young warrior's wrath [201-216]. 10) Priam tells her he is determined to enter the enemy camp [218-227]. 11) Zeus sends Hermes to protect Priam, [334-338]. 12) Hermes addresses Priam [460-467]. 13) Priam confronts Achilles [486-506]. 14) Achilles responds to Priam: "Do not grieve" [517-551]. 15) Priam expresses eagerness to take Hector's corpse [553-558]. 16) Achilles advises Priam to be patient, as a suppliant should [560-570]. 17) Achilles apologizes to

element in a contractual transaction; besides *offer* and *acceptance*, there must exist just compensation so that both contracting parties can seal a pact. [44] Hector's corpse becomes a metaphor for *proper consideration*; such is a possible early source for an essential element in civilized trade.[45]

We should become aware that the impending death of Achilles, prominent in the legend, is not a tragedy included in the *Iliad*, any more than the actual abduction of Helen is, nor is the fall of Troy. But by drawing a strong quadrangular equivalence between a) Agamemnon's quarrel with Achilles over Briseis, with the resulting loss of Patroclus, and b) the strife of Menelaus against Paris over Helen, with the subsequent death of Hector, the ransom for desire is redeemed by both sides. In this transposition Homer draws back from the *Iliad* scenario, with Achilles' wrath as central focus, to the cause for the War's inception in the legend due to the abduction of Helen. With the ransom of Hector's corpse, the issue of Achilles' wrath is finally resolved. Hector's sacrifice recalls Patroclus' death. In the bitterly ironic balance of war Hector had slayed Patroclus [*Il*. XVII, 821]. Priam's loss reminds Achilles of the loss his own father, Peleus, must also suffer when his son dies away from home. [46] Within confines of the epic narrative Homer does not let his choice of topic become either obscure or distant from the legendary themes. Hector and Patroclus as noble victims are not sacrificed in vain to blind desire. Priam and Achilles subdue the desperation

---

Patroclus' shade for giving back Hector's corpse [592-595]. 18) Lastly, Achilles tells Priam to share a meal before going back to Troy. [599-620]

[44] Fuller 8.

[45] There were requirements to be fulfilled to generate an obligation under *stipulatio* in Ancient Law. Principles of the *lex mercatoria* were received into the English Law of Contracts. Lord Mansfield held that "Mercantile law is not the law of a particular country, but the law of all nations." *Luke v. Lyde* (1759). *History of Contract Law*.en.wiki. In modern Anglo American strictly legal usage, we see that "Courts will not ask whether the thing which forms the consideration does in fact benefit the promisee, or a third party, or is of any substantial value to any one… Consideration means not so much that one party is profiting as that the other abandons some legal right in the present." Fuller 48-49. Priam kisses the hand that slew his son. Achilles puts up in his tent the King of Troy. No contract is drafted, but the pact is sealed.

[46] Priam reminds Achilles of his own father [*Il*. XXIV, 504]. And Achilles tenderly weeps thinking of Peleus [*Il*. XXIV, 511]. The lyric candor of this scene can hardly be matched in heroic narrative, for the great warrior weeps, not for his own death, but for the loss his father must subsequently endure.

engendered by a cruel destiny to wrench from grief a conceptual basis for the endurance of societal commerce. As symbol for *proper consideration* Hector's corpse leaves an indelible mark in the development of commercial trade in Western civilization. The concluding scene of the epic shows the burial of Hector's body, object of a peaceful transaction between sworn enemies. The profound significance establishes relevance for study of the Oral Epic beyond the realm of literature. At the end of the *Iliad* the frantic nature of heroic struggle merges into the legal framework for contractual exchange.

Works Cited

Adkins, A.W.H. *Moral Values and Political Behavior in Ancient Greece.* W.W. Norton, 1972.

Cunliffe, Richard J. *A Lexicon of the Homeric Dialect.* U of Oklahoma P, 1963.

Fuller, Lon L. and Eisenberg, Melvin A. *Basic Contract Law.* Thomson/West, 2006.

Gans, Eric. *Chronicles of Love and Resentment 771 Origins of GA II: The End of Culture?*
    https://anthropoetics.ucla.edu/view/vw771/2023.
    Signs of Paradox: Irony, Resentment, and Other Mimetic Structures. Stanford P, 1997.
    The End of Culture. U of California P, 1985.

Gierke, Otto. *Associations and Law.* Edited and translated by George Heiman. U of Toronto P,
    1977.

Girard, René. *Violence and the Sacred.* Translated by Patrick Gregory. The John Hopkins U P,
    1977.

Gordon, Cyrus H. *The Common Background of Greek and Hebrew Civilizations.* W.W. Norton,
    1965.

History of Contract Law, Wikipedia, The Free Encyclopedia, Wikipedia Foundation, 23
    October 2004,
    https://en.wikipedia.org.wiki/history_of_contract_law

Homer. *The Iliad.* https://www.perseus.tufts.edu/hopper/text?doc=Perseus:text:1999.01.0133

Jaeger, Werner. *Paideia: The Ideals of Greek Culture.* Vol. 1. Translated by Gilbert
    Highet. Oxford U P, 1970.

Lidell, Henry G. and Scott, Robert. *A Greek-English Lexicon.* Oxford, 1968.

Smyth, Herbert Weir. *Greek Grammar.* Benediction Classics, 2010.

The Companion Bible. Notes and Appendixes by E.W. Bullinger. Kregel Publications, 2004.

The Iliad of Homer. Translated by Richmond Lattimore. The U of Chicago P, 1951.

The Odyssey of Homer, Translated by R. Lattimore. Harper Perennial, 1991.

Voegelin, Eric. *Order in History.* Vol. 2. *The World of the Polis.* Louisiana State U P, 1957.

The Heroic Code and the Challenge
of Time in *Beowulf*

The epic poem *Beowulf* begins and ends with a burial scene. The hero is praised at death. Treasures are placed in the ship that bears Scyld Scefing's corpse, as he asked, *swa he selfa bæd* [29].[1] Similarly, a great hoard of wealth is piled on Beowulf's funeral pyre. The respect earned is proclaimed at the end of the poem in words which become emblems that recall the memory of a dead hero, kindest, *mildust* [3181], and most eager for praise, *lofgeornost* [3182]. [2] Glory depends on the performance of heroic deeds. The monarch earns the right to govern and counts on support from his indebted vassals. Consequently, governance is aligned to the giving of gifts. The throne is *gifstol* [168], the palace is *gifhealle* [838]; the monarch is a giver of treasures, *beaggyfan, sincgyfan, goldgyfan* [1102, 1342, 2652]. Pervasive use of "*gyfan*" compounds stresses the duties of king and subject. The heroic code depends on the combination of rights and corresponding duties. [3] Syntactic structures reflect the lexical expressions. Conceptual semantic analysis of the hypothetical statements reveals that the balance of rights and duties depends on the exercise of social, religious and political intentions. The boast before an antagonist becomes a challenge; a warrior's prayer for victory turns into a curse for the enemy, and an oath of loyalty opens liability for betrayal from other members of a tribal society. Historical and legal implications of this study remain possibilities for further development and research.[4]

---

[1] Quotes from the original text are from *Klaeber's Beowulf*. Fulk 4.
[2] Fred C. Robinson remarks how *lofgeornost*, "most vain glorious,... occurs most often in homiletic discussions of the cardinal sins... the only documentation for the positive sense of the word, 'most eager to deserve praise,' is the last line in *Beowulf*." *Style* 81.
[3] In his *The Cultural World of 'Beowulf'* John M. Hill remarks how "Beowulf's world is one in which gift exchange and feud are central." Hill 85.
[4] M.T. Clanchy remarks in his *From Memory to Written Record* that: "The most difficult problem in the history of literacy is appreciating what preceded it." Clanchy 27. The first *Beowulf* manuscript dates back from the mid-eleventh century. Kiernan 20. Albert B. Lord

At the opening of the poem the poet asks the audience to heed exploits of bygone princes. The glory of the kings, "*þeodcyninga þrim*," is the performance of princes, "*hu þa aeþelingas ellen fremedon*" [2-3]. Such concern for posterity, imminent from the start, is explored subsequently in the tale of Scyld Scyfing who a) was picked up as castaway [6b-7a], b) found consolation for this misfortune [7b], and c) eventually overcame kingdoms which paid him tribute [9,11]. The warrior prince himself becomes the heroic king. By giving gifts the young prince earns support from men for times of need:

> So should a (young) man, in his father's hold,
> do good by giving splendid gifts,
> so that later in life his men stand by him
> in turn, his people fast by, when war comes.
> By glorious deeds in any tribe a man prospers [20-24].[5]

> *Swa sceal ge(ong) guma gode gewyrcean,*
> *fromun feohgiftum on faeder (bea)rme,*
> *þaet hine on ylde eft gewunigen*
> *wilgesiþas, þonne wig cume,*
> *leode gelaesten; lofdaedum sceal*
> *in maegþa gehwaere man geþeon.* [20-24a]

From the very beginning of the epic we see that the king, Scyld Scefing, answered the crisis of leaderless anarchy, "*aldorlease*" [15], and, while fulfilling the people's needs, he is justified in expecting support in return. Thus each party has the right to the promise of performance. The magnanimity of the prince who dispenses gifts is dependent on the promise of support by the subjects at a time of need. Since their society subsists among warring pre-feudal tribes, loyalty is tested by courage in battle. The phrase which introduces the six lines quoted above is "*Swa sceal ge(ong) guma*," a formula recurring with slight variation:

---

insists that "the technique of formulaic repetition proves that *Beowulf* was composed orally." Lord 198-199. Andy Orchard states that calculations for the poem's actual origin range in estimates from the seventh to the eleventh century. Orchard 6-7. Colin Chase mentions events in the poem which occurred before the mid-sixth century. Chase 3. We may consolidate these remarks by allowing for the existence of an oral tradition.

[5] Translations from the original text are mine.

1172b – when Queen Wealtheow asks Beowulf to give a discourse on friendship,
for so should a man do, while she hands him the drinking cup [1169];

1534b – then in the *scop's* description of a Beowulf fearless against Grendel's
dam, and ready to trade life for glory, as becomes a man;

2166b – also in Beowulf's speech to Hygelac about Hrothgar's recompense to
him before leaving his court, as befits a kinsman;

2708 – evidently, as Wiglaf's assistance proves vital in the final encounter. So a
thane should be to a lord in need;

3174 – and, finally, at the end of the epic, in mourning a dead king it is fitting
for a man to speak praise.

These scenes of giving and acceptance show that remembrance is meaningful; yet,
among our examples, the most gruesome encounters, the Beowulf-Grendel's dam
duel and the hero's death in the final battle, show that expectation itself can prove
useless because the future is unpredictable.

The oral epics in the Western World at times share parallel schemes of
symbolic representation. The prevalent influence over different cultures across
geographic borders is *Homer*. [6] In order to achieve proper interpretation for the
plot of *Beowulf* we recall the three roles for the Homeric god which Hugh Lloyd-
Jones outlines in his *The Justice of Zeus*. Zeus Xeinios is the protector of guests
and hosts; Zeus Hikesios is the god of suppliants, and Zeus Horkios is the lord
presiding over oaths. [7] The functions of the divinity are derived from the words
for guest, *xenos*, suppliant, *hikete*, and oath, *horkos*. These three categories reveal
respectively a social, religious, and political perspective. Since in the Oral Epic
the reader encounters a society of warriors, the three socio-cultural functions for
the role of the divinity are manifested in human behavior as boast, prayer and
oath. In his *Structural Anthropology* Claude Lévi-Strauss has pointed out how

---

[6] Lord draws a number of parallels between *The Song of Roland*, the *Iliad*, and *Beowulf*. He
compares Aeschere to Patroclus, "the friend who is killed before the encounter of the hero with
the enemy." Lord 201. About Marsile and Grendel, Lord states that they both suffer similar
wounds, the loss of an arm; and "they both seek solace from a female, Grendel from his dam,
Marsile from his wife." Lord 206-207. In general form, Lord also compares *Beowulf* to *Homer*
in regards to "repeated assemblies with speeches, repetition of journeying from one place to
another, and on the larger canvas the repeated multiform scenes of the slaying of monsters."
Lord 198-199.

[7] Lloyd-Jones 5.

opposing binary structures reflect tribal customs in a primitive society.[8] Consequently, we may view the three linguistic categories boast, prayer, and oath, as including their three corresponding adverse counterparts, challenge, curse, and betrayal. The most basic oscillation between the positive and the negative meaning for each thematic role may shift, depending on the speech situation. For instance, a boast at the mead hall remains a boast, but when formulated at a scene of confrontation with an antagonist, the boast becomes a challenge. Likewise, a hero's prayer for victory is a curse for his opponent. Finally, formulating an oath of alliance opens the possibility for breach of trust, or cruel betrayal. In the military context of the Oral Epic, we perceive that double possibilities within each role expand plot motivation and character development. We refer to the serpentine movement between positive and negative social, religious, and political behavior along the plot of *Beowulf* as the heroic code.

The conditional statement with the *gif* adversative conjunction is found in the hero's first recognition scene. Beowulf, having heard of Grendel's deeds, leaves the fold to secure future alliance for his king. When the coastguard of the Scyldings asks the Geats for identification, in order to grant right of passage, he recognizes the hero as outstanding, unless his countenance belies him, *naefne him his wlite leoge* [250]. The hero, in turn, asks the warrior to judge his knowledge of affairs: "you know if it is, as we truly hear say." [9] The *gif* clause [272b] introduces ravages perpetrated by the "mysterious persecutor," *deogol daedhata* [275]. The actual remedy proffered follows suit:

> I may offer Hrothgar,
> through boundless spirit, good advice,
> how he, old and good, may overcome his enemy,
> if reversal should ever come to him,
> remedy from evil trouble.
>
> *Ic þaes Hroðgar maeg*
> *þurh rumne sefan raed gelaeran,*

---

[8] Lévi-Strauss 161.
[9] *þu wast, gif hit is / swa we soþlice secgan hyrdon,* [272b-273]

áu he frod ond god feond overswydeb –
gyf him ed-venden aefre scolde
bealuwa bisigu bot eft cuman – [277b-281]

After asking the coastguard whether report of the crisis is true, the hero alleges that he can advise Hrothgar how change may be brought about if relief is to come. Since Beowulf and his men are strangers, the statement is crucial and daring enough when considered as means for armed men without password, *leafnesword* [245b], to gain safe passage. The noun *edwenden*, reversal, is in apposition to *bot*, remedy; the reversal implied, to be acknowledged as remedy, is the very relief to be attained through projected success of the mission led by Beowulf. Should help be refused, chances are Hrothgar may suffer sad necessity as long as Heorot stands. Although the hero's boast allows for failure, since he does not say that he will surely succeed, the immediate responsibility for the decision to grant safe passage to the troop of sailors rests with the coastguard.

The burden on the sentinel is heavy. He responds with an aphorism:

Each of us both shall
know a sharp shield bearer,
from his words and deeds, he who thinks well.

Aeghwaebres sceal
scearp scyldwiga gescad witan,
worda ond worca, se be wel benced. [287b-289]

The phrase *worda ond worca* is a universalizing doublet. This combination of two dissimilar terms which complement each other becomes a metaphor for the decision to maintain the spoken word as a man's intention to perform accordingly. We agree with Stanley B. Greenfield that the maxim discriminates between words and deeds, and does not combine both in the province of heroic prowess, since Beowulf has not yet performed. [10] A coastguard should know the difference between someone who lies, and one who speaks truthfully. Words have strong

---

[10] Greenfield, "Of Words and Deeds," 49-50.

significance in human action when spoken by a man who keeps them. [11] As an inclusive metaphor with general significance in the poem the principal symbolism of the doublet becomes entwined in context with the development of a boast to vow progression; the latter is heroic consequence of the former. The tendency to insure that words are kept draws in the poem a relation between axiom and experience; provisions must be made for the elements of risk. In his monumental grammar Otto Jespersen defines a conditional statement as "the preparation for a possible contingency." [12] Beowulf's original boast was knowledge of dreadful events and belief that he could remedy the situation at Heorot. Diplomatic tact is the norm when the hero speaks his name to a nobleman, *Beowulf is min nama* [343]. He will help the king if Hrothgar allows the heroes to enter his court: "if he will grant us / that we may greet his kind self." [13] The boast is toned down in order to further his mission inland and reach the crown.

Disorder in the realm remains pervasive even among the ruling class. Unferth, a counselor at Hrothgar's court suggests in a notorious digression that, since Beowulf had lost once a swim match to Breca, he may lose if he seeks an encounter with Grendel:

> I expect from you a worse settlement,
> although you have availed yourself everywhere in battle,
> hard struggle, if you dare wait
> for Grendel close-by all night long.

---

[11] Beowulf insists that his state of mind is loyal, *holdne hige* [267a]. Discussing meaning of the term, John M. Hill cites D.H. Green, who "has pointed out the reciprocal and even legal nature of the word *hold* as an ethical term within the *comitatus* and as a term for oaths and contracts, where the *huldi* (protection) of the lord is involved as a guarantee for truth of the statement." Hill 71. Safe passage is granted because the troop can be considered a loyal group, *hold weorod*. [290b]

[12] Jespersen 367. Since we have mentioned kennings with the verb *gifan*, and propose to analyze phrases introduced by the *gif* conjunction, a philological note is in order. The adversative adverbial conjunction comes from Old Gothic *jabai*; whereas the verb *gifan* stems from the Indo-European root **ghab*, which gave Latin *capio* and *habere*. Barney 20, 62. Fernand Mossé in his *Manuel de la langue gotique* explains uses in Old Gothic for the adversative adverbial conjunction *jabai* in hypothetical statements. Mossé 198. There is no morphological connection between *gif* and *gifan;* the "y" is used interchangeably with "i" in the textual script of both terms with no apparent distinction in sense or voicing. Fulk 385, 389.

[13] *gif he us geunnan wile/ þaet we hine swa godne gretan moton.* [346b-347]

Ðonne weᵢ̃e ic to be wyrsan gebingea,
deah bu headoraesa gehwaer dohte,
grimre gude, gif bu Grendles dearst
nihtlongne fyrst nean bidan. [525-528]

Unferth's challenge suggests that the "attempted settlement with Grendel will be worse than what was dealt to Beowulf in that foolish episode with Breca." [14] The hero justifies his loss to Breca in the swim match by revealing his strife with water monsters along the way. To answer fully the impudent challenge, Beowulf must unveil the root of decadence in the governing body. The hero then proceeds to identify Unferth as killer of his own brother, *beah du binum brodrum to banan wurde,* [587]; and adds that Grendel would not attack Heorot if the nobleman's military stamina would match his words:

Never had Grendel caused so much slaughter,
fiendish monster, to your lord,
humiliation at Heorot, if your mind were,
your spirit in battle, as fierce as you suppose.

baet naefre Gre[n]del swa fela gryra gefremede,
atol aeglaecc ealdre binum,
hyndo on Heorote, gif bin hige waere,
sefa swa searogrim, swa bu self talast; [591-594]

The hero's response to the insult relates the etiology for disorder at Heorot directly to the envious Unferth. [15] In his *Cain and Beowulf* David Williams remarks that: "the prediction of the Fall of Heorot through parricide and the stories, for instance, of Heremod and Unferth extend the image of a socially ever present Cain to the past and future of the Danes, yoking time and space in the universalization of the theme." [16] Grendel as allegorical figure, descendent of Cain, is a projection of a

---

[14] Hill 78.

[15] "Envy was seen as pre-eminent in Cain's motive for murder, as it had been in Satan's temptation of Eve, and the hatred borne by Cain for Abel was seen as the envious hatred by the evil of the good simply because they are good." Williams 23. Through borrowed Old Testament symbolism archetypal enemies sprang from Cain. [107-108; 1261-1263]

[16] Williams 41.

deep rooted social evil which apparently afflicts the royal court. Slaying of a brother is an over-looming symbol for high treason. Beowulf's reproach to Unferth suggests that through decadence at the royal court a counselor's advice may degenerate into a demonic curse. Hrothgar's thane fails to discern a correspondence between spoken words and actual prowess.

We are now in a position to account for the Finn Lay's relevance in the plot of the epic. The story tells how Queen Hildeburh loses her brother, Hnaef, when Frisians battle Danes; Finn himself, the lady's husband, is later inevitably slain. She was of Danish stock and her spouse was Frisian. Hengest, after taking over command, is persuaded to abide by a treaty whereby Finn's men cannot mention to the Danes their equivocal position. The crisis the treaty purports to solve is the bitter slaughter of kinfolk, *moðorbealo maga* [1079], a most serious offense. The Frisians swear to uphold the peace, *geðingo budon* [1085b], provided that there is no grudge on either side. Both tribes should receive an equal share of the treasures, *feohgyftum* [1089]. Spanning prohibition, the treaty includes a penalty clause:

> They confirmed by treaty on both sides
> a firm peace. Finn declared to Hengest
> undisputed zeal by oaths,
> that the survivors, by decision of the wise,
> held fast with honor, so that any man
> who would break the truce by word or deed,
> or through enmity ever complain,
> though now they follow their ring giver's murderer,
> deprived of chief, what to them was imposed,
> that if any Frisian by audacious speech
> were to bring remembrance of the murder-hatred,
> then the edge of the sword should settle it.

> *Ða hie getruwedon on twa healfa*
> *faeste frioðuwaere. Fin Hengeste*
> *elne unflitme aðum benemde;*
> *þaet he þa wealafe weotena dome*
> *arum heolde, þaet ðaer aenig mon*
> *wordum ne worcum waere ne braece,*
> *ne þurh inwitsearo aefre gemaenden,*

ðeah hie hira beaggyfan banan folgedon
ðeodenlease, þa him swa geþearfod waes;
gyf þonne Frysna hwylc frecnan spraece
ðaes moþorhetes myndgiend waere,
þonne hit swordes ecg syððan scede. [1095-1106]

Both sides are bound to the peace compact, *frioðuwaere*. Discussing the meaning of *getruwian* in Old English, D.H.Green explains that "the verb can be employed, as in *Beowulf*, to denote the formal conclusion of a treaty between two tribes, or, in strictly legal literature, in the sense of proving one's innocence and clearing oneself from a legal charge." [17] Finn, with a dwindled host, swears by oath, *aðum benemde*, to hold in honor the Danes and kill any man on either side who would break the truce by word or deed, *wordum ne worcum* [1100]. [18] But the Danes resent their equivocal position as thanes loyal to the slayer of their leader; and, eventually, Finn, who never reaches a suitable bargaining position, is killed; then Hildeburh, Finn's widow and Hnaef's sister, is taken to Denmark [1057-1059a]. The penalty clause included in the edict does not ameliorate Finn's downfall, rather, it is a sign that disturbance will recur. As Fabienne Michelet notes: "Of course, these fragile arrangements fail." [19] The negative elements of the boast-

---

[17] Green 251. The philologist quotes from the compilation of Anglo Saxon laws collected by Felix Liebermann: "If he (a person accused of plotting against his lord) wishes to make himself trustworthy, let him offer the king's *wergild*, payment for his life." Liebermann 50. See also Bosworth: *Gif he hine selfne triowan wille, do ðaet be cyninges wergelde*. Bosworth 1014. Green cites a variant for the same characteristic conditional statement, found in legal texts, mentioning that *treowsian* can have a meaning identical to *triowan*: *Gyf he hine sylfne treowsian wille*. Liebermann 51. Evidently, to pledge guarantee is assumed to establish loyalty and attain enduring trust.

[18] The formula shows Anglo Saxon legal usage. The phrase appears in the characteristic Colyton oath. An indentured servant should be trusted to speak the truth when swearing loyalty. The swearer vows to his lord to be "loyal and true... and never of one's own will or power, in words or deeds, do anything that to him is hateful," *hold and getriwe... and naefre willes ne gewealdes, wordes ne weorces owiht dom þaes him labre biþ*." Hill 71. Liebermann cites the oath as prevalent between 920 and 1050 A.D. Liebermann 396. The use in *Beowulf* reveals that the universalizing doublet existed centuries before in oral culture.

[19] Michelet comments on lines 1099b-1103. Michelet 36. The universalizing doublet proves to no avail. We should also note that the present participle, *myndgiend* [1105], "reminding," used with the verb "to be," has a ponderously lasting effect. Frederick Klaeber compares the use here to the effect of Grendel's act of constantly lying in wait, *ehtende* [159b]. Fulk 125, 370, 415. Combined with *frecnan spraece*, the prohibitory clause has a foreboding tone.

vow progression, a challenge which, through curse, ends in betrayal, prevail over the penalty clause.[20] In the context of the *Beowulf* epic, the Finn Lay exemplifies the case of an oath which fails to secure permanent loyalty. The slaying of a brother in law is a metaphor for high treason and reminds us of the crisis without peaceful remedy wrought in Heorot by the descendant of Cain, Grendel.[21]

In contrast, Beowulf exhibits loyalty, courage, and strength. He is so strong that he can afford courtesy. When Beowulf announces his mission to Hrothgar our hero wishes to display, along with valor, the discretion that will endear him to his lord back home, Hygelac. The boast is tempered by caution when he weights the grim balance of war before facing Grendel:

> I expect he will, if he may reach control
> in the wine hall, eat unafraid
> the Geat people, as he oft did
> the glorious host. Never need you
> bury my head, for he will devour
> my slaughtered corpse, if death takes me;

---

[20] The opportunity for ransom in *The Battle of Maldon* is equally insulting. The invaders frame their request for ransom in two *gif* clauses:

> It is not necessary that we destroy each other.
> If you are good for the proper amount,
> we will confirm peace with the gold,
> if you decide who here is most powerful,
> so that you will ransom your people,
> pay seafarers, according to their choice,
> money for peace, and take a truce from us.

> *Ne burfe we us spillan gif ge spedab to bam;*
> *we willab wib bam golde grib faestnian.*
> *Gif bu baet geraedest, be her ricost eart,*
> *baet bu bine leode liesan wille,*
> *sellan sae-mannum on hira selfra dom*
> *feoh wib freode and niman frib aet us.*[34-39]

Byrthnoth interprets the proposal as a challenge, and addresses the men with the famous line: "they will give to you spears as tribute," *Hie willab eow to gafole garas sellan.* [46] Pope 17.
[21] Andrew Barton reminds us that the characters themselves are not aware of Grendel's connection to Cain [1355]. In a poem that provides detailed lineage for its characters, Hrothgar's and Beowulf's ignorance about Grendel's descent makes the fiend appear even more incomprehensibly hideous. Barton 14.

54

nor bear away the slain he likes to savor;
solitary, he eats without mourning,
stains his moor retreat; never you need
to sorrow much over my corpse.
Send to Hygelac, if battle takes me,
the best of mail shirts which guards my breast.

Wen' ic þæt he wille, gif he wealdan mot,
in þaem guðsele Geatena leode
etan unforhte, swa he oft dyde,
maegenhreð manna. Na þu mine þearft
hafalan hydan, ac he me habban wile
d'r]eore fahne, gif mec dead nimeð;
byreð blodig wael, byrgean þenceð,
eteð angenga unmurnlice,
mearcað morhopu; no ðu ymb mines ne þearft
lices feorme leng sorgian.
Onsend Higelace, gif mec hild nime,
beaduscruda betst, þaet mine breost wered, [442-453]

The repetition of *gif* clauses stresses the stern pessimism of three potential consequences ensuing from Beowulf's hypothetical defeat: a) the monster will continue feeding at Heorot, b) the beast's cannibalism obviates the need for burial, and c) the hero's byrnie should go to Hygelac as Hrothgar's last respects to a dead champion. This triple *gif* clause speech exposes the negative side of the encounter's outcome. The fiend's opponent presents a boast expressing deep regret. Grendel has eaten Danes before; precedent dictates that he may eat Geats too, if he prevails in the encounter. Denial of burial represents the tragic curse of a cruel fate, if death takes the hero. Beowulf then closes the initial boast with the sad final request to honor the alliance between Geats and Danes: his armor should go back to Hygelac if the hero dies in battle. [22] The request includes the hope that the feud against Grendel does not break the Geat-Dane alliance. Submission to an uncontrollable fate funnels choices into a personal decision. Courage is mastered

---

[22] Ursula Schaefer cites lines 452-453a, as an example of a clause that splits "with verb and direct object," reading: "Send to Hygelac (if battle takes me) the best of mail shirts." Schaefer 112. To take the *gif* clause as strictly parenthetical seems to detract from the rhetorical impact, although the syntactical notion Schaefer remarks certainly applies.

by a subdued but mature will since the speech opens and closes with acceptance of the ineffability of destiny "he shall trust / the judgment of God, he who death takes" [440b-441b]; [23] and then follows the aphorism: "fate always goes as it shall" [455b]. [24] The tone of the speech is to let the audience ponder over whether a boast could approximate a prayer by observing the possibility for failure.

In the plot, Hrothgar, rising to the occasion, promises by oath to his guest that if he survives he will lack nothing: "There will be no lack of reward for you / if you survive this courageous deed with life" [660b-661]. [25] Before the actual encounter Beowulf proclaims that he shall fight Grendel without weapons, since the monster scorns them, a boast tempered by honor:

> but we both in the night shall
> forsake swords, if he dares to seek
> war without weapon; and then let wise God
> to whichever side grant the glory,
> as is deemed proper.

> *ac wit on niht sculon*
> *secge ofersittan, gif he gesecean dear*
> *wig ofer waepen, ond sibðan witig God*
> *on swa hwaebere hond halig dryhten*
> *maerðo deme, swa him gemet bince.* [683-687]

We may consider a positive meaning for the hypothesis of unarmed struggle: if the fiend dares to seek battle without weapon they will fight, and the balance of an even match shall be tilted by God alone, a boast well within the prayer framework. [26]

Hope in the wake of a harsh fate is expressed by Wealtheow during celebrations after Grendel's defeat. In a double *gif* clause speech the queen tells

---

[23] *ðaer gelyfan sceal / dryhtnes dome se be hine dead nimeð.* [440b-441b]

[24] *Gaeð a wyrd swa hio scel!* [455b]

[25] *Ne bið be wilna gad / gif bu baet ellenweorc aldre gedigest.* [660b-661]

[26] As it turns out, the hero tears off Grendel's arm and the fatally wounded fiend runs in anguish to his lair [815b-822]. So the subdued challenge to an apparent wrestling match becomes the terrible curse of a slow death for the enemy.

Hrothgar that, if the king should die before his own son comes of age, then his younger brother's son, Hrothulf, shall rule the young men with honor [1182b]. The king's nephew should repay prince Hrethric with kindness if he remembers all graces bestowed on him while growing up at court [1185a]. [27] Wealtheow caps her view of fate with the hope that loyalty may keep the kingdom together. Again we sense in the use of the hypothetical phrase a desire to provide hope that a personal decision projected into the future may alter the fatal outcome of an uncontrollable fate. A sincere intention and the proper remembrance of past favors meet the challenge of an unknown destiny. [28]

As the plot unravels the Grendel adventure brings temporary success. Although the hero is confident to have restored order in Heorot, nevertheless, Grendel's dam attacks to avenge her son. She wreaks havoc at Heorot. The ravages include the loss of Aeschere, Hrothgar's dear counselor. Vile turmoil at Heorot affects the high echelons of society at the royal court. Subsequently, the *scop* quotes Beowulf in indirect discourse to outline the hero's trusting incredulity of a crisis he believed ended. We are told the hero asks Hrothar, in a presumptuous boast, "if the night had been / agreeable according to his desire" [1319b-1320]. [29] The negative response shows Beowulf that, sadly, the relief obtained by heroic prowess does not endure. Hrothgar henceforth expounds on the revenge of Grendel's mother and ends his speech with a request for Beowulf:

---

[27] *Ic mine can / glaedne Hrobulf, þaet he þa geogode wile / arum healdan, gyf þu aer þonne he, / wine Scildinga, worold oflaetest; / wene ic þaet he mid gode gyldan wille / uncran eaferan, gif he þaet eal gemon, / umborwesendum aer arna gefremedon.* [1180b-1187]

[28] In a poem with allegorical creatures and mythical heroes we do not expect correspondence to historical truth. However, we note the curious fact that: "From Scandinavian sources it is known that after Hrothgar's death Hrothulf usurped the Danish throne, and killed Hrethric, Hrothgar's son and heir." Wright 120. The Queen's optimistic expectations are betrayed in real time by a cruel fate.

[29] *fraegn gif him waere / aefter neodladum niht getaese* [1319b-1320]. This is one of only two occasions in which the *scop* uses the *gif* clause himself in the text of the epic; we consider this instance akin to the character's protracted dialogue since it is a direct reference to Beowulf's thought, not an extrinsic description, as occurs in 2841a. In the second instance the *scop* intrudes on the narrative by assigning to the hero a measure of *hybris* (cf. infra, footnote # 41). Even in the first instance here, there is an implied criticism of excessive naiveté. In retrospect, the *scop's* use of his own unique *gif* clauses contain both times reproaches to Beowulf in indirect discourse.

seek if you dare!
I shall reward you for this battle with riches,
with ancient treasures as I did before,
twisted gold, if you come back.

sec gif þu dyrre!
Ic þe þa faehđe feo leanige,
ealdgestreonum, swa ic aer dyde,
wundnam golde, gyf þu on weg cymest. [1379b-1382]

The king pleads with the hero, as he did before [660b-661], to risk his life if he dares, and gifts will follow, if he returns alive. Hrothgar's vow again shows the promise of reward for specific performance. Beowulf answers with an equivalent set of double *gif* clauses. The hero repeats his previous request to Hrothgar in parallel contractual fashion:

if I at your service should
lose my life, you be to me,
when dead, in the position of a father.
Be you protector to my young retainers,
close companions, if battle takes me.

gif ic aet þearfe þinre scolde
aldre linnan þaet đu me a waere
forđgewitenum on faeder staele.
Wes þu mundbora minum magoþegnum,
hondgesellum, gif mec hild nime. [1477-1481]

Half-line 1481b is identical to 452b.[30] The text expresses the recurring echo of a deal struck anew. Starting with the coastguard scene, and on to the scene of departure from Horthgar's court, closure for speeches negotiating promises is provided by a characteristic hypothetical statement.

After the successful encounter with Grendel's mother, involving submersion and subsequent rebirth from deep waters, Beowulf prepares to sail. In

---

[30] Fulk 17, 51.

the hero's farewell speech to Hothgar the non-truth conditional statement solidifies into a semantic framework for an oath of contractual alliance:

If I may earn in any way
on earth your affection further,
Lord of men, than I have yet
with war prowess, I am prepared immediately.
If I hear over the expanse of the waves
that a nearby terror threatens you
as these enemies formerly did,
I shall bring a thousand thanes,
warriors as help. I know from Hygelac,
lord of the Geats, though he be a young
guardian of the people, that he will further me
in words and deeds, since I praise you well,
and I should bear the spear shaft as help for you,
military support, when you need men.
If thence Hrethric to the court of the Geats
should betake himself, as child of princes, he may there
find many friends, distant lands
are better sought by he who is strong himself.

Gif ic bonne on eorban owihte maeg
binre modufan maran tilian,
gumena dryhten, donne ic gyt dyde,
gudgeweorca, ic beo gearo sona.
Gif ic baet gefricge ofer floda begang,
baet bec yrbsittend egesan bywad,
swa bec hetende hwilum dydon,
ic de busenda begna bringe,
haeleba to helpe. Ic on Higelace wat,
Geata dryhten, beah de he geong sy,
folces hyrde, baet he mec fremman wile
wordum ona worcum baet ic be wel herige
ond be to geoce garholt bere,
maegenes fultum, baer de bid manna bearf.
Gif him bonne Hrebric to hofum Geata
gebinged beodnes bearn, he maeg baer fela
freonda findan; feorcybde beod
selran gesohte baem be him selfa deah. [1822-1839]

In this leave-taking *gif* clauses head three lines outlining a sealed pact. The first *gif* clause is the hero's pledge of future service, *guðgeweorca* [1825], if he hears about trouble from abroad while at home. The second clause outlines the promise of additional reinforcements from overseas in case of need, the hero's own reliance on his king; Hygelac proffers aid according to his usual pledge by words and deeds, *wordum ond worcum* [1833]. The doublet conferring desert of safe-passage for the troop through conviction of good faith, used in the coastguard scene [289a], expresses reliance on a promise; its use here again represents firm trust on human intention. Beowulf closes the speech with an oath of future alliance and a concern for posterity in the form of an invitation extended to Hrethric to visit Hygelac's court and find support as suppliant in adversity.

Appearing regularly in speeches framing contractual relations, the conditional statements introduced by the *gif* particle represent a character's commitment to the development of events in the plot. Beowulf is a character prominent enough to include a sequence of three consecutive *gif* clauses in two of his speeches [cf. 442b, 447b, 452b & cf. 1822a, 1826a, 1836a]. A notable difference between these two triple "*gif* clause" speeches is that in the first instance all three apodoses precede the protases, outlining a twinge of fatalism which could be interpreted as humility on the part of the hero. Stanley B. Greenfield finds tripartite subsequent responses by Hrothgar to both Grendel adventures:

> "... reward for such favors rendered is obviously implicit in the heroic ethos, and Hrothgar makes it explicit when he finally accepts Beowulf's offer [660-661]. When Beowulf performs as promised, Hrothgar not only wishes to adopt him as a son, but again states that he will give him treasure [946b-950]; and at the banquet that evening he does so, the poet describing the gifts in detail [1020-1049]. The same pattern holds in the Grendel mother's episode: promise before action in 1730-1732, promise after action in 1706b-1707a, and 1783b-1784, and actual giving in 1866-1867." [31]

---

[31] Gifstol 109.

In plot development, the king answers the boast-vow progression in kind. But, whereas Beowulf's speech preceding the Grendel encounter contains, with ascending pessimism, provisions for valiant death, the later farewell speech strikes a more hopeful note: in the wake of need, Beowulf will come with one thousand men as aid and Hrethic will find friends at Hygelac's court, if he decides to go, *gebinced*. [1837]

Hrothgar replies with recognition of the hero's unrivaled integrity. There is no wiser young warrior [1842c-1845a]. Beowulf can bargain with words, *bingian*. The recognition of the hero has progressed considerably. Now he is the mature speaker who can settle disputes. Decorous rhetoric is vital to tribes engaged in constant warfare since an alliance is maintained through the promise of support. Hrothgar seals the Danish farewell to Beowulf acknowledging that an uncontrollable fate should be dealt with through the strong commitment of a personal decision. The king recognizes that the hero's maturity is essential for the Geats:

> I consider the fact
> that, if it happens that the spear take,
> in fierce battle, the son of Hrethel,
> your lord, or either illness or steel,
> guardian of the people, and you are still alive,
> the Seafaring Geats do not have
> any better king to choose, to be
> guardian of treasure for heroes, if you would rule
> the kingdom for kinsmen.

> *Wen ic talige,*
> *gif baet gegangeð baet ðe gar nymeð,*
> *hild heorugrimme Hrebles eaferan,*
> *adl obðe iren ealdor ðinne,*
> *folces hyrde, ond bu bin feorh hafast,*
> *baet be Sae-Geatas selran naebben*
> *to geceosenne cyning aenigne,*
> *hordweard haeleba, gyf bu healdan wylt*
> *maga rice.* [1845b-1853a]

The double *gif* clause speech discloses tension between an unpredictable destiny splayed broadly in future time and an indomitable human will: a) if your king should be killed and you still live, I expect the Geats will choose you king; that is, b) if you decide to rule your people as guardian of wealth for heroes. The first *protasis* postulates the death of Hygelac. Beowulf should be the next king upon death of the monarch. The Geats would certainly choose an appropriate giver of treasure, if Beowulf should undertake the task to rule over their dominion. The second *apodosis* is subdued; the term "king," *cyning* [1851], is in apposition to "guardian of the treasure for heroes," *hordweard haeleþa* [1852]. We may heighten the strength of this suggested *apodosis* with the verb "to be" by viewing *haleþa* as an objective, rather than as a partitive, genitive; since clearly, grammatically, it may be both. Beowulf as king is a guardian for the heroes' wealth, not just a keeper of treasure. Hrothgar presents a hopeful prayer, and seals the course of destiny with the possibility of a pact to assure the hero's future. A cruel fate can be dealt with through a bright prospect. The prowess of Beowulf as hero, and the uncompromising gratitude of Hrothgar as ruler, bring the Geats and Danes to remain reconciled in brotherhood. [1855-1857]

The contractual nature of the conditional statements, oscillating from boast through prayer and on to oath in the speeches of characters, indicates a keen appreciation for human intentions on the part of the poet. However, upon return to Hygelac's court, absence of dramatic dialogue accounts for scant *gif* clauses. Historical narrative in indirect discourse predominates in the last part of the epic. Beowulf's relationship with his king is perfect and needs no further development in confrontational dialogue. A conditional statement, after all, is a pre-empted provision for the future dependent on behavioral contingencies unfulfilled in the present. Nevertheless, there is dramatic tension in plot development because in the poem "nothing can stand except in negation," as J. D. Niles says:

> The dominant mood created by this recurrent
> play of joy against sorrow, creation against dissolution,
> may strike some readers as fatalistic, and it may well be;

but, if so, the poem's fatalism stems from a realistic understanding of the limits that bound earthly success. The *Beowulf*
poet seems to have lived enough of life to appreciate the awful
ease with which time and an indifferent fate blot out even the
most glorious of human achievements. [32]

We do not know how Beowulf, who reigns for two lines, has trouble after fifty
years [2207-2211]. [33] The unpredictable quality of human life engulfs Beowulf as
an absolute contingency.

Although, as he approaches his last adventure, Beowulf is guiltless, the
aged king ransacks his soul to find cause for the dragon crisis: "the wise man
believed that he had angered bitterly the eternal creator God, over an old law"
[2329-2331a]. [34] He, nevertheless, relies on the experience of past exploits to
confront the new evil. Yet the hero is now an old man and his boast has weakened.
The *scop* tells how the hero speaks boasting words for the last time: *beotwordum
spraec / niehstan siðe* [2510b-2511a]. *Beot*, a word often translated as boast, also
means promise.[35] Beowulf wishes to draw the dragon from his lair:

---

[32] Niles 932.

[33] Division of the poem is a complex issue. According to the XIXth century view of multiple
lays, championed by Karl Müllenhoff, there are four sections: the Grendel adventure (1-836),
the Grendel dam encounter (837-1628), Beowulf's homecoming (1629-2199), and the final
Dragon episode (2200-3182). Shippey 155. Thomas A. Shippy winds his way through several
issues by quoting the views of J.R.R. Tolkien, Adrien Bonjour, and Arthur Brodeur. Tolkien
divides the poem into two parts: A (1-2199), the voyage, and B (2200-3182), the return. Both
textual critics, Müllenhoff and Tolkien, break the text as Beowulf assumes kingship. Bonjour
supports the poem's unity by asserting that the episodes and digressions are signs of a poet
working by irony and contrast. Brodeur points out that, if the poet had concentrated on historical
heroes over mythical monsters, "we should have gained a kind of English *chanson de geste*,
and lost the world's noblest *Heldenleben*." Shippey 163. James Earl suggests that the shift from
hero to *king* might have been perceived by an audience of "warrior-aristocrats" as a shift from
ego to superego, a projection of inner inadequacy. Shippey 173. Regarding the issue of various
segments, Lord states: "the fact that these parts *might* or *could* be sung separately would not
mitigate against *Beowulf* as a single song." Lord 200. We should consider that a peaceful reign
that lasts two lines, with a drastic and sudden political turnover after fifty years, shows the
desperate efforts of a *scop* trying to provide closure for an oral story spanning over several
centuries.

[34] *wende se wisa, þaet he wealderde / ofer ealde riht ecean dryhtne / bitre gebulge,* [2330-
2331]

[35] Klaeber 306.

I waged many
battles in youth, yet I,
old folk guard, shall seek a fight,
perform a glorious deed, if this evil ravager
out of his earth hall attacks me.

'*Ic genðde fela*
guða on geogoðe; gyt ic wille,
frod folces weard faehðe secan,
maerðu fremman, gif mec se mansceaða
of eorðsele ut geseceð.' [2511b-2515]

Apparently, old age is not an acceptable justification for inaction; he must perform as before; but there is a melancholic tone in the challenge. We sense that the hero expects that the antagonist may not leave his abode. If the dragon comes out, he shall engage him. The present general condition presents a lamentable probability. The tone of regret deepens as the hero gives up his emblematic grip:

I would not bear a sword,
a weapon for a serpent, if I knew how
to contend with might against this dragon,
to grapple bravely, as I did formerly with Grendel.

'*Nolde ic sweord beran,*
waepen to wyrme, gif ic wiste hu
wið ðam aglaecean elles meahte
gylpe wiðgripan, swa ic gio wið Grendle dyde; [2518b-2521]

In this conditional statement the *apodosis* prominently precedes the *protasis*. Curiously enough, Beowulf is not sure whether he knows how to grapple boastfully, *gylpe wiðgripan* [2521a]. Both clauses are in the subjunctive mood.[36] The notorious should-would condition becomes a would-should hypothesis.: "I would forsake weapons if I knew (or "should know") how to grapple with the monster according to my boast." He does not know how to grapple, so he will not

---

[36] Fulk 457, 458. Alan Renoir states that: "The clause *gif ic wiste* gives the utterance an unmistakable tone of hesitancy, and there is a suggestion of nostalgia in the reference to the fight with Grendel." Renoir 247.

forsake weapons. The great leader is wavering: "I am courageous, so I shall forego boasting against this war flyer" [2527b-2528]. [37] By skipping the boast altogether Beowulf trusts destiny. This encounter is his last adventure. The hero commands his warriors to stand back as he claims to be alone the dragon's antagonist. He tells the men in his troop:

> This is not your venture,
> not set for any man, except for me alone
> against this monster, to undertake with strength,
> perform as nobleman.

> *Nis þaet eower sið,*
> *ne gemet mannes, nefn(e) min anes,*
> *þaet he wið aglaecean eofoðo daele,*
> *eorlscype efne.* [2532b-2535a]

The exception outlined by the conjunction, *nefne* [2533b], is crucial, for courage is the trademark of Beowulf as warrior king, his uniqueness. Having attained old age in an unforgiving world is a fact which confirms the tested strength of his courageous nature, although his stamina may now be deficient. Despite the presumptuous absence of a boast, the sensitive reader may consider the hero's stance boastful enough; we should remark, that, since the confrontation is close at hand, and the dragon is near, the boast is direct enough to be a challenge.

The hero pays dearly for his daring. At the time of most need only one warrior is torn with sorrow. As kinsman, Wiglaf means well, *þam ðe wel þenceð* [2601b]. This phrase recalls the coastguard's aphoristic reply, *se þe wel þynceð* [289b]. Harboring good intentions is the trait of the dependable warrior. The noble retainer reminds the other warriors of the oath sworn:

> I remember the time when we partook of mead,
> when we promised our lord,
> in the great hall, who gave us these rings,
> that we would repay him for this war gear,

---

[37] *Ic eom on mode from, / þaet ic wið þone guðlogan gylp ofersitte.* [2527b-2528]

if to him such need arose,
for helmets and hard swords.

Ic ðaet mael geman, þaer we medu þegun,
þonne we geheton ussum hlaforde
in biorsele, ðe us ðas beagas geaf,
þaet we him ða guðgetawa gyldan woldon,
gif him þyslicu þearf gelumpe,
helmas ond heard sweord. [2633-2637]

The optative mood of *gelumpe* and *gyldan woldon* signals a contrary to fact condition: "We swore to help if there were need." [38] The condition precedent to the warriors' promise for performance is no longer hypothetical. The impending circumstances press for the ostensible response – i.e. there is need, so we must help. Since the hero is struck down, aid from his retainers is demanded by the situation, regardless of his previous command for them to stand back [2529]. Wiglaf tries to persuade the men that following specific orders to the letter could become an excuse for disobedience. There is need of courage to face danger by undertaking fulfillment of the promise to act. Although Beowulf thought to do it alone [2643], he needs friends, *maegenes behofað* [2647b]. Circumstances have changed. The issue of desertion becomes semantic when the loyal warrior reminds the others of their duty to perform in case of need. He who gave treasure considered the men worthy of decision-making. The order to remain at bay is not inflexible; nor is it an excuse to ratify cowardice.

Wiglaf stands apart as the exceptional, dutiful thane. The death of the dragon itself is a joint venture between Beowulf and Wiglaf [2706-2708a]. Wiglaf's difference from the deserters is that he answers with courage the call of loyalty at a time of need. He is the proverbial thane at hand in time of need, *þegn aet ðearfe* [2709a]. Twice he remarks that he would rather die with his gold-giver than live with shame.[39] The deserters' failure to keep their oath of allegiance by

---

[38] Klaeber 369, 425.

[39] Wiglaf says: "It is much better for me that fire / engulf my body along with my gold-giver." *þaet me is micle leofre, þaet mine lichaman/ mid minne goldgyfan gled faeðmie* [2651-2652];

performance at a time of need makes them liars as well as cowards, paradigms of treason, "the ten weak traitors together" *tydre treowlogan tyne aetsomne* [2847]. The pronounced alliteration strikes a note of profound chastisement and sticks to the reader's ear and memory. Consequently, a tragic drama develops in the course of the last adventure. After the monster's fatal attack, the dying hero's stance before the dragon's hoard is prefaced by the line, "the old man contemplates in grief the gold" [2793]. [40] The *scop*, who has restrained in general from using *gif* clauses himself directly in the narration, presents a delayed *parainesis*:

> Indeed, I have heard said
> that on land there is hardly a man,
> among heroes who are victorious,
> although he were daring in action,
> such as to be able to rush a poison breather,
> or stir with his hands a hoard,
> if he found the guard awake,
> dwelling in the barrow.

> *Huru þaet on land lyt manna ðah*
> *maegenagendra mine gefraege,*
> *þech ðe he daeda gehwaes dystig waere,*
> *þaet he wið attorsceaðan orede geraesde,*
> *oðde hringsele hondum styrede,*
> *gif he waeccende weard onfunde*
> *buon on beorge.* [2836-2842a]

There is an implication of *hybris* or *démesure* on the part of the hero since he approached the gigantic worm when the creature was awake, whereas the fugitive slave who caused the crisis succeeded in plundering the hoard while the monster was sleeping [2290]. The *scop* uses his own conditional statement for censure. Beowulf's defiant attitude seems reckless.

In a world of uncertainty, the persistent use of conditional statements reflects the belief that judgment should be tailored to different situations with

---

and again, later: " Death is better / for every warrior than a shameful life." *Dead bið sella / eorla gehwylcum þonne edwitlif!* [2890a-2891]
[40] *gomel on giohðe, gold sceawode.* [2793]

altered consequences. There was need for the coastguard to be persuaded that orders are not binding in all cases, so as to grant Beowulf's troop the right to have safe passage [280]; and this same claim is Wiglaf's implied reproach to the deserters [2637]. We should uphold the spirit and not just the letter of the law. Beowulf had to insist that the seafaring Geats must penetrate Danish territory; Wiglaf had to insist that the troop of men not hang back despite the leader's orders. The former is a subdued boast, the latter scolds the warriors for shameful cowardice. Both poles of attempts at persuasion, the noble boast and the stern rebuke for treason, are, respectively, the first and last passages where the prominent *gif* clause is employed by the characters themselves. Use of the hypothetical statement starts in the epic with a positive boast and winds its way through challenge, prayer, curse, and oath to end in a reprimand for deplorable betrayal. Subsequently, we perceive that the *scop's* second use of the conditional clause in the complete course of the actual narrative suggests that the hero is rash in his old age. [41] Expressed in non-truth conditional semantics, the heroic code allows the sensitive reader to view the complete epic plot of *Beowulf*.

We endeavor to conclude. The conceptual notion of balance and exchange, seen already as part of the semantic structure in the first three quarters of the epic poem, becomes in the last part pervaded with bitter irony. The dragon's hoard, buried by the last living member of an extinct clan, is a legacy to the earth men could not enjoy [2247-2466]. The despairing elegiac tone of the ending broadcasts the theme of waste after generations of splendor, the betrayal of time. Consequently, a monstrous creature guards the hoard which will profit him in no way, *ne byð him wihte ðy sel* [2277b]. The futility of riches is brought into focus because the gold transferred from the dragon's hoard into Beowulf's barrow at his burial will be as useless to others as it was before, *eldum swa unnyt, swa hyt (aero)r waes* [3168]. Yet, poignantly, the hero thanks god before death for the treasure he sought to acquire for his people [2797-2798]. Beowulf had boasted

---

[41] The only other use of the *gif* clause by the *scop* quotes in indirect statement Beowulf's incredulous inquiry after the visit of Grendel's dam [1319b-1320] (cf. supra, footnote # 29).

earlier that Hygelac needed to buy, *gecypan* [2496], no better warrior than him to ride point on the war trail. Beowulf paid in war what Hygelac granted him; as he says: *geald aet gude, swa me gifede waes* [2491]. The quest for glorious victory brings anecdotal defeat. The dragon made plunder of the hoard no easy bargain, *yde ceap* [2415], for anyone. The word *ceap* hardly alters in meaning as used in Beowulf's autobiographical speech [2426-2509].[42] In this context we find that Haethcyn, after his accidental killing of Herebeald, his older brother, ascended to the throne and waged war against neighboring tribes. [43] Haethcyn himself paid for a raid against Swedes with his own life, a high price, *heardan ceap* [2482a]. Mercantile language acquires lugubrious overtones in the text of *Beowulf.* Although wealth is a symbol for honor, the messenger at the end explains in his elegy how all the grimly purchased hoard, *grimme geceapod* [3012b], shall burn and be buried along with Beowulf. The hoard was dearly bought, *deade forgolden* [2843b], and grimly begotten, as granted by a harsh destiny, *grimme gegongen; waes baet gifede to swid* [3085]. Expressions for trade and commerce define wasteful prowess in the last part of the epic.

The hero takes up his last stand as ultimate liability for lordship over his dominions. His open challenge to the gigantic worm ends any possibility to sue for peace *freode to friclan* [2556]. On the other hand, the uncondemned slave, *unfaege* [2291], plunders under the auspice of fate, not the heroic code. Fleeing a whipping for wrong doing, in need of shelter, the slave steals a cup. The pernicious theft ends Beowulf's reign of fifty years as well as the dragon's rest of three centuries [2278-2279]. Both unforeseen contingencies are introduced each

---

[42] Fulk 360.

[43] Both, Herebeald and Haethcyn, brothers of Hygelac, are sons of Hrethel, Beowulf's maternal grandfather. Stefan Jarisinski argues that *feohleas* [2441], in reference to the accidental slaying of the heir to the throne, "does not mean *without remedy*, but *remedied only by death*." Jarisinski 115. The unintentional nature of the hunting accident draws sympathy from the modern reader. In Late Antiquity, the epic poem's contemporary audience probably felt intense grief at recollection of an event which must have stirred tragic catharsis in a cultural milieu that had "no fully developed legal vocabulary for negligence." Jarisinski 117. Furthermore, in Germanic belief, Odin may insist on avenging slain kin. Jarisinski 144. Therefore, Beowulf's eventual ascension to the throne, after his maternal uncle Haethcyn's death in battle, may have been viewed by the *scop's* audience as residing under cloudy augurs.

by an *oðdaet* clause [cf. 2210b & 2280b]. Beowulf rules "until a monster plunders in the night." [44] The death shadow guards the hoard "until an intruder enraged him." [45] Thus, the chain of causation consists of a dragon usurping the peace after Beowulf's fifty year reign because a fugitive slave steals a cup to give as peace offering to his lord. The irony is sullen and acute. The stolen cup becomes legal restitution, or peace offering, *frioðowaere* [2282], which the slave's master accepts as trade for a granted pardon, *bene getiðad* [2284b]. Hence the issue of a private bargain for redemption leads to an accursed end. Neither Beowulf nor the plundering slave can be considered blameworthy. The Grendel episodes epitomize the betrayal of kinsmen which engenders endless feuds; the dragon encounter symbolizes the progress by destruction of a society advanced enough to work metal with fire. Wiglaf explains how Beowulf could not be persuaded to hold back from the fiery dragon by any counsel [3079-3081]. He met fire with steel. [46] Little fault can be found in the hero except adherence to a rigid heroic code even in old age.

Beowulf's sacrifice leads him to attain a new sacred dimension. Wiglaf is the one who reprimands the deserters. We sense that Beowulf's heroic stature does not allow him to descend to the level necessary to address deserters. Fred C. Robinson remarks: "At his death, Beowulf never condemns the cowardly retainers who deserted him in his hour of need; his thoughts are always and exclusively on the survival of his people." [47] Robinson insists that Beowulf's nobility goes beyond what is normally assigned to mortals. The textual critic from Yale indicates that the cremation of the dead hero, along with the burial of lavish obsequies alongside his ashes [3137-3148], together with the procession of princes chanting dirges around his gravemound [3169-3182], hint at deification.

---

[44] *oðdaet an ongan / deorcum nihtum draca ricsian.* [2210b-2211b]
[45] *oðdaet hyne an abealch / mon on mode.*[2280b-2281a]
[46] Beowulf ordered an iron shield made for him, because linden wood could not withstand the dragon's fire. [2337-2341]
[47] *Tomb* 65.

[48] Robinson mentions further that fallen pagan heroes were often venerated at a *heroön*, a shrine located at the tomb. [49] The funeral of Scyld Scefing at the opening of the poem is not followed by horsemen circling a barrow.[50] Beowulf's funeral rites are more extensive; the monument built around the burial site seems redundant, something like a second funeral [3155-3160]. [51] The landmark should serve as a beacon for seafarers sailing the seas, as the hero himself requests at death [2802-2808]. [52] Beowulf's burial represents the last rites earned by a dutiful monarch.

The detailed progression of language as a means to reflect proper exchange, seen through the rhetorical use of conditional statements in the first part of the poem, falls away into paradox and adulation in the last part. The ensuing bafflement results in a dramatic tension which must have thrilled the poet's audience. Yet the hero of innate good intentions remains true to his word to the very end; [53] then he meets his judgment day, *domes daeg* [3069]. The language for exchange reveals in its development from logic to ironic paradox hopes which are betrayed eventually by time. The heroic code seeks to overcome social evils by providing expectation of rights for duties among noble thanes. Unfortunately, even when there is great strength of character, an uncertain fate overcomes the exceptional hero. The insistent use of language to express vows for projected heroic action, which drive the leader to a sacrificial death, sounds the plaint of a pagan civilization at the brink of Christianity. The death of Beowulf is the end of

---

[48] *Tomb* 11.

[49] *Tomb* 7.

[50] Gale R. Owen-Crocker identifies cremation of Hnaef and the remains of warriors from both sides in the Finn Lay [1107-1113]. There is also a funeral suggested in the Last Survivor's lament. Owen-Crocker 45-48, 61, 147-148. But the two main funerals in the text are Scyld Scefing's and Beowulf's.

[51] *Tomb* 17.

[52] Robinson suggests that the high monument built as beacon to guide distant sailors discloses a clear connection with Christ: "The noun *becn* [3160] means 'sign, portent, idol,' and it is used in Christian times to refer to the Cross and to Christ's miracles." *Tomb* 17-18.

[53] Beowulf states proudly in his last words that he did not swear falsely, "I did not swear many oaths wrongfully," *ne me swor fela / aða on unrihte* [2738b-2739a]. S.O. Andrews remarks how "*not many* is a frequent litotes for *not at all*." Andrews 95.

an age. We can recapture lost ideals through interpretative analysis of a poem which contains a conceptual transition in the notion of exchange that evolves from a rudimentary social contract on to the promise of specific performance within the sworn word, regardless of the outcome. The *Beowulf* poet strives to conform the vows for display of heroic action to a balance of rights for duties; yet the progress from boast to oath is strained by the equivalence of an analogous opposing binary tension from challenge, through curse, to betrayal because, in the Anglo Saxon society of the *Beowulf* epic, positivist thought is tempered by a fatalistic view of temporal reality.

Works Cited

Andrews, S.O. *Postcript on "Beowulf."* Russell & Russell, 1969.

Barney, Stephen A. *Word-Hoard. An Introduction to Old English Vocabulary.* Yale UP, 1977.

Barton, Andrew. *The Knight's Progress and Virtual Realities: The Medieval Adventure from Beowulf to Ready Player One.* 2018. Texas State U, MA thesis.

Beowulf. Translated by David Wright. Penguin Books, 1968.

Bosworth, Joseph. *An Anglo-Saxon Dictionary.* Edited by T. Northcote Toller. Oxford UP, 1976.

Chase, Colin. "Opinions on the Date of *Beowulf*, 1815-1980." *The Dating of Beowulf*, edited by Colin Chase. U of Toronto, 1981. pp. 3-8.

Clanchy, M.T. *From Memory to Written Record: England, 1066-1307.* Harvard UP. 1979.

Fulk, R.D., et al., editors. *Klaeber's Beowulf and the Fight at Finnsburg.* 4th ed. U of Toronto P, 2008.

Green, D.H. *The Carolingian Lord: Semantic Studies in Four Old High German Words: Balder, Frô, Truhtin, Hêrro.* Cambridge UP, 1965.

Greenfield, Stanley B. *"Gifstol* and *goldhoard* in *Beowulf." Old English Studies in Honor of John C. Pope,* edited by Robert B. Burlin and Edward B. Irving, Jr. U of Toronto, 1974, pp. 107-117.

------------ "Of Words and Deeds: the Coastguard's Maxim Once More." *The Wisdom of Poetry,* edited by Larry D. Benson and Siegfried Wenzel. Western Michigan U, 1982, pp. 45-51.

Hill, John M. *The Cultural World of 'Beowulf.'* U of Toronto P, 1965.

Jespersen, O. *A Modern English Grammar.* Vol. 5. A. Stifsbogtrykkeri, 1949.

Jarisinski, Stefan. *Ancient Priviliges: "Beowulf," Law, and the Making of Germanic Antiquity.* West Virginia UP. 2006.

Kiernan, Kevin S. "The Eleventh-Century Origin of *Beowulf* and the *Beowulf* Manuscript." *The Dating of Beowulf*, edited by Colin Chase. U of Toronto, 1981, pp. 9-21.

Klaeber, Fr., editor. *Beowulf and the Fight at Finnsburg.* 3rd ed. D.C. Heath, 1950.

Liebermann, Felix, editor. *Die Gesetze der Angelsachsen.* Vol. 1. S.M. Niemeyer, 1916.

Lévi-Strauss, Claude. *Structural Anthropology.* Translated by Claire Jacobson and Brook Grundfest Schoepf, Basic Book, 1963.

Lloyd-Jones, Hugh. *The Justice of Zeus.* U of California P, 1983.

Lord, Albert B. *The Singer of Tales.* Atheneum, 1976.

Michelet, Fabienne L. "Hospitality, Hostility, and Peacemaking in *Beowulf." Philological Quarterly,* vol 94, nos. 1/2, 2015, pp. 23-50.

Mossé, Fernand. *Manuel de la langue gotique.* Aubier, 1956.

Niles, J.D. "Ring Composition and the Structure of Beowulf." PMLA, vol 94, no. 4, 1974, pp. 924-935.

Orchard, Andy. *A Critical Companion to "Beowulf."* D.S. Brewer, 2007.

Owen-Crocker, Gale R. *The Four Funerals in Beowulf.* Manchester UP, 2009.

Pope, John C., editor. *Seven Old English Poems.* Bobbs-Merrill, 1976.

Renoir, Alain. "The Heroic Oath in *Beowulf,* the *Chanson de Roland,* And the *Nibelungenlied." Studies in Old English Literature in Honor of Arthur G. Brodeur,* edited by Stanley B Greenfield. U of Oregon Books, 1963, pp. 237-267.

Robinson, Fred C. *"Beowulf" and the Appositive Style.* U of Tennessee P, 1985.

------------ *The Tomb of Beowulf and Other Essays on Old English.* Blackwell Publishers, 1993.

Schaefer, Ursula. "Rhetoric and Style." *A Beowulf Handbook,* edited by Robert E. Bjork and John D. Niles. U of Nebraska P, 1997, pp. 105-124.

Shippey, Thomas A. "Structure and Unity." *A Beowulf Handbook,* edited by Robert E. Bjork and John D. Niles. U of Nebraska P, 1997, pp. 149-174.

Williams, David. *Cain and "Beowulf:" A Study in Secular Allegory*. U of Toronto P, 1982.

The Oliphant and Roland's Sacrificial Death

To wage war does not necessarily include the willingness to be martyred. Yet strong devotion to a cause may result in resigned submission to death. Desire for victory may incline warriors in the first *Crusades* toward sacrifice. The *chansons de geste* explore such feelings. The poems are set in the early IXth century, but they were conceived in their present form in the XIIth century. In between these centuries the transition from the Carolingian to the Capetian kings marks a historical shift in France from monarchy to vassalage. The king retained the throne, but lost control of the land.[1] Perhaps for this reason the epics show a longing for a bygone era.

At times heroes encounter deceit and pursue a course of action in a role marginal to the acceptable cultural setting, such as William's betrayal by monks in the *Moniage Guillaume*.[2] A missed reward in the *Charroi de Nîmes* leads the hero to accept disguise as a battle tactic; and he even accepts the geographic move to another area of the kingdom in *La Prise d'Orange*.[3] In the *Guillaume* cycle

---

[1] Ferrante 23.

[2] Ferrante 21, 288. In the context of this epic, *William in the Monastery*, the hero retires to a monastery, where he experiences the monks' hypocrisy in flagrant breaches over their own rules of silence and attendance at services. In turn, William's uncouth habits and large size appall the monks who set him up to be robbed and killed by thieves; but the hero wins the fight. Eventually learning of the ruse, the disappointed hero leaves the monastery and becomes a hermit.

[3] Gégou 14-15. Ferrante 16, 18. The epics of the *Guillaume* cycle are interconnected. In *The Crowning of Louis* William travels back and forth from Paris to Rome, constantly saving King Louis and the Pope from enemies. Since the King forgets William when he distributes gifts at the beginning of the *Charroi*, in the *Couronnement* Louis offers lands to the hero outright; but the offered lands have heirs already. Incensed, William refuses to take them and decides instead to conquer lands held by the Saracens, first Nîmes, and then Orange. To enter Nîmes, William disguises himself. His skill speaking Arabic comes in handy. Although he eventually is successful, he and his relatives become restless and leave for Orange. In *The Conquest of Orange* William feels especially enticed by possession of the magnificent palace and the beautiful queen Orable. Again disguised, William, with his brother Gilbert, and his nephew Guielin, enter the Gloriette castle, but they are eventually discovered. William sends Gilbert to bring Bertrand, his other brother, with help. After a stint in towers and subterranean passages,

William repeatedly experiences a measure of dishonor and misery. But the sacrifice of William in the *Couronnement de Louis*, serving a backward king, son of Charlemagne, seems slight compared to the sacrifice of Roland in the *Song* that bears his name.[4]

The Song of Roland has a greater tone of seriousness in contrast to other *chansons de geste*. The *Chanson* shares the rather naïve view of the struggle between Christians and pagans common to the genre. Saracens are portrayed as morally deficient. To battle Charlemagne, the emir relies on treachery; and the pagans' chief counsel, Blancandrin, even advises sending their sons as guarantors for a false oath to Charles, knowing full well that they may lose their lives.[5] The Christian life is generally confined to belief in angels, use of relics, and spirited fulfillment of vows that often culminate in boastful oaths. However, there are long prayers before military encounters in *Aliscans* and the *Couronnement* which are not present in the *Roland* epic.[6] William flees when odds are stacked against him in battle and lives to fight another day.[7] Such behavior would be unheard of in our *Roland*. The ambush at Roncevaux makes *The Song of Roland* more tragic.

The higher level of suspense and intrigue is seen most notably in the rapport between characters. There is charm in the fantastic Rainoart's relation to William

---

William is successful. Aided by Bertrand, he wins the city and marries Orable, who is baptized and takes the name Guiborc.

[4] Ferrante 33-35, 66.

[5] Terry 4-5.

[6] Ferrante 26, 32, 83-85, 91-92.

[7] Ferrante 40. Bennett 93, 97. *La Chanson de Guillaume* presents a series of battles. Vivien, William's nephew, is mortally wounded, defending a hopeless position, in the first battle. William hears about the battle at Archamp and leaves home with his own nephew Girard, and his wife's nephew Guishard. Both youngsters die in the second battle and William leaves the battlefield to return Guishard's body to Guiborc, his wife. At home he is treated to a hearty meal and goes back to battle with reinforcements. Back at the third battle, William is victorious against the Saracen Derame, whom he slays with help from his nephew Gui, renowned for his small stature. William then finds the dying Vivien and gives him communion before he dies. William returns home and is not recognized by his wife until he shows her the famous wound to his nose. She urges him to seek aid from the king. He then leaves for the royal court. There is intrigue at Louis's court. William's father, Nemeri of Nerbune, smoothes things down. After a huge feast, the army marches south joined by the powerful Rainoart, former kitchen helper. In the fourth battle the pagans are caught picnicking, so victory is swift. After further intrigue at court, Rainoart's identity as Guiborc's brother is revealed. The poem ends on this note.

as comrade.[8] We experience sorrow in *Aliscans*, and in the *Chanson de Guillaume* when we grieve over William's encounter with his dying nephew, Vivien;[9] but the melodrama of the hero who pulls through every tight spot throughout the remaining of the plot, together with the feasting ludicrously integrated into battle scenes, projects an image of protagonists lacking the intercourse we witness in Oliver and Roland's remarkable personality conflict.[10] The characters' goal in the epic genre remains to defeat the enemy. But we sense that the *jongleur* of *The Song of Roland* has a more complex agenda than the oral poets in other *chansons de geste*. Roland's sacrifice becomes more important than the attainment of victory as crucial factor essential to establish a primordial example for the best military leader. The uniqueness of *The Song of Roland* accounts for its endurance as the most popular *chanson de geste* of the Medieval Era.

In his *Originary Thinking: Elements of Generative Anthropology*, Eric Gans mentions the concept of a "neo-classical esthetic," where the protagonist is aware of his separation from the scene of representation in which he must find significance.[11] Roland's basic instinct is to pursue glory; this desire drives him irremissibly toward Roncevaux, a scene of *ostensive*, self-confirming sacrifice. In contrast, Oliver grasps only the worldly *imperative* need to kill the enemy and win the war. Hence as one character's behavior points out the central significance of Roncevaux, the other wishes to disengage himself from absorbed entanglement in a hopeless predicament. The measure of Roland's valor is his pursuit of glory; and essential to such quest is his delay in sounding the Oliphant. The hero's sacrifice in turn imparts courage to others.

This deferral of action delays resolution and increases violence in *The Song of Roland*. The battle at Roncevaux delays the eventual downfall of the Saracen

---

[8] Ferrante 36. *Aliscans*, another name for Archamp, tells basically the same story as *La Chanson de Guillaume*, except that the death of Vivien opens the poem, instead of extended battles, these come later; in addition, Rainoart's deeds are recounted in more detail.

[9] Ferrante 19, 204. Bennett 128-131.

[10] Ferrante 193. Bennett 85, 102.

[11] Originary Thinking 151.

Army. Ambush through Ganelon's treason provides the necessary motivation for plot development. But it is within the battle itself that we witness the key source for the final tragedy of the French rearguard: Roland's refusal to sound the Oliphant in time. Roland's delay provides the necessary deferral, after the rearguard's separation from the Frankish host, to turn the Oliphant's sound into an *ostensive* call that broadcasts awareness of events without requesting action, so that only later will Charlemagne and his Army avenge the hero's sacrificial death. Since Roland's heroism subsequently inspires warriors in Charlemagne's Army, the hero's sacrifice provides a crucial role model for future warfare. By loosing a battle, the claim to glory becomes mounted on sacrifice. In *The Song of Roland* the main character's martyrdom provides enough dramatic content and significant justification to prolong the war. Deferral of victory is vital for development of the epic narrative.

In terms of military strategy, Oliver's plea to sound the Oliphant is based on the obvious superiority in numbers of the Saracen host. By contrast, the French are few: "It seems to me, our Franks are very few!"[1050][12] Roland transposes the issue to an idealistic plane; the reason the hero gives for not sounding the Oliphant, when Oliver first asks, is fear of suffering from lack of *glory*, Medieval French *los*, in the future:

> "Roland, my friend, it's time to sound your horn,
> King Charles will hear, and bring his army back."
> Roland replies, "You must think I've gone mad!
> In all sweet France I'd forfeit my good name."
> ....
> *"Cumpaign Rollant, kar sunez vostre corn,*
> *Si l'orrat Carles, si returnerat l'ost."*
> *Respunt Rollant: "Jo fereie que fols!*
> *En dulce France en perdreie mun los."*[1051-1054]

---

[12] Terry 43. Line references and quotes in English are from the Patricia Terry translation, unless otherwise specified. In general, Terry's translation seems to follow the Gérard Moignet edition. All quotes from the Medieval French are from the Moignet edition. *"De nos Franceis m'i semblet aveir mult poi!"*[1050] Moignet 94.

"Good name" is here the equivalent of *los*. As we shall see, the glorious name, or *los*, Roland is keen to protect will depend on his readiness to make the ultimate sacrifice, and on the delayed *ostensive* use of the horn in order to call attention to his stance in the sacrificial scene. This absolute need requires that he not use the Oliphant primarily as an *imperative* signal first; such obstinacy makes it impossible for other agents, Charles and his Army, to prevent him from assuming his heroic role. No one can devalue the sacrificial scene of Roncevaux. The decision not to sound the horn at the instant in the plot when timely arrival of reinforcements seems possible distinguishes the two heroes.

A second time, Oliver reiterates his plea [1059-1060], only to get another justification for delay from the valiant chevalier:

> Roland replies, "Almighty God forbid
> That I bring shame upon my family,
> And cause sweet France to fall into disgrace!
> I'll strike that horde with my good Durendal."
> ....
> *Respont Rollant: "Ne placet Damnedeu*
> *Que mi parent pur mei seient blasmet*
> *Ne France dulce ja cheet en viltet!*
> *Einz i ferrai de Durendal asez."* [1062-1065]

Roland abhors the thought that his lineage be tarnished by lack of courage. The hero intends to use his sword, *Durendal*, not the Oliphant.[13] Brandishing his sword is preferable to blowing the horn. Roland is not one to let a call for help take precedence over the warrior spirit:

> "No man on earth shall have the right to say
> That I for pagans sounded the Oliphant!"
> ....
> *"Que ço seit dit de nul hume vivant,*
> *Ne par paien, que ja seie cornant!"* [1074-1075]

---

[13] *Durendal* was given to Roland directly by Charlemagne, perhaps in the knighthood ceremony. *Durendal ,/ Ma bonne espee, que li reis me dunat* [1120-1121]. Farnsworth 48, 54.

No warrior on either side could blame Roland for seeking aid by sounding the horn too early. His intention is to achieve Eternal Glory, not safeguard mortal life. Salvation is in Heaven. In the context of the *Chanson*, the hero's example of faith will give transcendental significance to his martyrdom within the epic world. In this sense, Roland turns his death into a scene of sacred representation.

As Oliver enters his third and final plea, trying in vain to get Roland to sound the horn, the same reason of uneven sides is brought forth: "Our company numbers but very few "[1087].[14] Roland at this point makes no reference to fear of losing glorious honor, *los*, but rather crowns his refusal by posing a direct comparative to the superiority in numbers of the Saracen army. He insists that vis à vis the huge host, his arduous courage is a superior force: "The better then we'll fight!"[1088][15] This could make Roland seem guilty of *hybris*, for refusing to sound the Oliphant in time. In his rebuttal, however, the brave hero insists that France should not lose worth:

> "If it please God and His angelic host,
> I won't betray the glory of sweet France!
> Better to die than learn to live with shame-
> Charles loves us more as our keen swords win fame."
> ....
> "*Ne place Damnedeu ne ses angles*
> *Que ja pur mei perdet sa valur France!*
> *Melz voeill murir que huntage me venget.*
> *Pur ben ferir l'emperere plus nos aimet.*"[1089-1092]

We should pose, in counterpoint to excessive courage, a concern for displaying, through self sacrifice, strict adherence to raw duty: a true warrior must remain at his assigned post to fullfil his duty, regardless of the consequences.

---

[14] "*Nus i avum mult petite cumpaigne.*"[1087]
[15] The actual Medieval French reads: "*Mis talenz en est graigne,*" literally, "My ardor is superior to that."(trans. mine) In his edition, Moignet notes: "*Graigne*, comparative form of *granz*; the usual form for the nominative case is, in Medieval French, *graindre* < *grandior*; the form *graigne* is reconstructed analogically from the oblique case *graignur* < *grandiorem.*" (trans. mine) Moignet 96.

When Oliver seems resigned to accept Roland's refusal to sound the horn, after his third plea goes unanswered, his tone becomes mournful. He remarks that the rearguard is worthy of pity, "who would not pity them!"[1104][16] Roland answers this observation declaring his sworn duty to remain in his post as leader of the regiment guarding Charles' return to France: "We'll hold our ground; if they will meet us here" [1103].[17] A vassal must safeguard his lord at all costs in the spirit of sacrifice [1009].[18] Submission to the warrior code is prominent in Roland's stance. To hold the rearguard in place becomes a strong mandate, as Oliver himself declares, even after perceiving the vast superiority in numbers:

> "Frenchmen, my lords, now God give you the strength
> To stand your ground, and keep us from defeat."
> ....
> *"Seignurs Franceis, de Deu aiez vertut!*
> *El camp estez, que ne seium vencuz!"* [1045-1046]

By remaining firm in its post the rearguard exemplifies absolute loyalty to the king with blind patriotic zeal. Both peers' loyalty to Charlemagne also brings out Ganelon's opposite role as traitor and instigator of the tragic events in the plot. Ganelon's later trial and execution broadcasts Roland's death beyond Roncevaux and into epic myth by juxtaposing extreme loyalty to absolute treason, negating the supremacy of self gain over patriotic zeal.

Roland's sacrificial stance is expressed by the hero twice:

> "This is the service a vassal owes his Lord:
> To suffer hardships, endure great heat and cold,
> And in battle to lose both hair and hide."
> ---
> "In his Lord's service a man must suffer pain,
> Bitterest cold and burning heat endure;
> He must be willing to lose his flesh and blood."

---

[16] *"Veeir poez, dolente est la rereguarde."* [1104]

[17] *"Nus remeindrum en estal en la place."* [1108]

[18] The text reads: *"Ben devuns ci estre pur nostre rei"* [1009], literally: "We must hold ground here for our king."(trans mine) Moignet 92.

....
"Pur sun seignor deit hom susfrir destreiz
E endurer e granz chalz e granz freiz,
Sin deit hom perdre e del quir e del peil."[1010-1012]

"Purs sun seignur deit hum susfrir granz mals
E endurer e forz freiz e granz chalz,
Sin deit hom perdre del sanc e de la char."[1117-1119]

A warrior must endure suffering for his king. The only situation that might prevent total sacrifice becomes the arrival of reinforcements. We should recall how, once the circumstances made aid futile, Oliver himself accepts the consequences for failing to sound the Oliphant in time. Practical necessity requires a readiness to fight, as the enemy closes in from all sides:

> "They are very close, the king too far away.
> You were too proud to sound the Oliphant:
> If Charles were with us we would not come to grief.
> Look up above us, close to the Gate of Spain."
> ....
> *"Cist nus sunt prés, mais trop nus est loinz Carles.*
> *Vostre olifan, suner vos nel deignastes;*
> *Fust i li reis, n'i oüssum damage.*
> *Guardez amunt devers les porz d'Espaigne."*[1100-1103]

Oliver's loyalty is not affected by his regret over the king's absence.[19] Oliver and Roland both agree that they must stand their ground. The warrior-bishop Turpin latches on to the quest for glory through sacrificial martyrdom:

> "My noble lords, Charlemagne left us here,
> And may our deaths do honor to the king.
> Now you must help defend our holy Faith!"[1127-1129]
> ---
> "Confess your sins, ask God to pardon you;
> I'll grant you absolution to save your souls.
> Your deaths would be a holy martyrdom."[1132-1134]

---

[19] The Terry translation calls Roland "proud." The original text states literally: "You did not deign sound the Oliphant;" (trans. mine).

....

"Seignurs baruns, Carles nus laissat ci;
Pur nostre rei devum nus ben murir.
Chrestïentet aidez a sustenir!"[1127-1129]

"Clamez vos culpes, si preiez Deu mercit;
Asoldrai vos pur voz anmes guarir.
Se vos murez, esterez sinz martirs."[1132-1134]

Joseph Bédier quotes line 1134 in his *Commentaires* to assert that the *jongleurs*, singers of tales, plied their trade on the by ways along medieval pilgrimage routes.[20] We concur that propagation of the Faith is meaningful enough, so that such contention holds sway at the most tragic moment of the *Chanson*, during the destruction of the rearguard; for the theme of warfare for Christ must exist within a specific socio-cultural milieu.[21] The present belongs to king and country; the future belongs to God.

The hero's obstinate insistence not to sound the Oliphant at the time reinforcements could save the rearguard turns him and his men into sacrificial victims, patriots willing to die for sacred ideals. It is only later on in the narrative, when approaching his end at the fierce battle in Roncevaux, that Roland finally acquiesces, "I'll blow my horn, and Charlemagne will hear" [1714].[22] Charlemagne hears, but he cannot arrive in time. Instead of calling the Army back, the Oliphant announces the rearguard's martyrdom. The issue of avoiding danger is no longer in question. Surrounded, the heroes are close to death. Oliver reverses his plea to blow the Oliphant because the instrumental cause for the call is gone:

Oliver says, "Then you'll disgrace your name.
Each time I asked you, companion, you refused.
If Charles were with us, we would not come to grief."

....

Dist Oliver "Ne sereit vasselage!
Quant jel vos dis, cumpainz, vos ne deignastes.

---

[20] Bédier 17.
[21] The route to *Santiago de Compostela*, a popular medieval pilgrimage site, passed by Roncevaux to continue on through to Northern Spain. Jotischky & Hull 61.
[22] *"Jo cornerai, si l'orrat li reis Karles."* [1714]

Were Charles and his royal Army present, there would be no defeat. The hour is late. The futility of the effort in effect invalidates the companion's reason in the original plea, repeated thrice [1051,1059,1071]. Regardless of vast superiority in numbers, a resistant stance is the only choice.

After the battle rages on fiercely for hundreds of lines, Turpin finally intervenes. The warrior-bishop brings to a conclusion the Roland-Oliver conflict about failing to sound the Oliphant with a realistic view of the situation:

> "End your dispute, I pray you, in God's name.
> It's too late now to blow the horn for help,
> But just the same, that's what you'd better do.
> If the king comes, at least we'll be avenged."
> ....
> *"Pur Deu vos pri, ne vos cuntralïez!*
> *Ja li corners ne nos avreit mester,*
> *Mais nepurquant si est il asez melz:*
> *Venget li reis, si nus purrat venger."*[1741-1744]

Turpin's argument pairs the uselessness of the act, as death approaches, to the impossibility of Charlemagne's timely presence. Yet Charles' Army must arrive, however late, at the Oliphant's call, so that the corpses will be respectfully buried [1749-1751]. Their martyrdom, with Roland at the center, has to be recognized. This ostension, distant in time and space, broadcasts a vital revelation to the community at large in establishing the significance of the original sacrificial scene.[23]

When at long last Roland blows the Oliphant for the first time the horn's sound is loud and clear:

---

[23] Such scenic source may be contrasted with a warning cry of "Fire!" which points to a communal danger and the decision to flee immediately, or fight the fire subsequently. Origin 43, 46. In our *Roland* what is overtly designated is a sacred *locus*, Roncevaux. The esthetic effect is pronounced, as the close-up in a modern day cinematographic *zoom-in* secures enthralled participation by the audience.

Count Roland presses the horn against his mouth;
He grasps it hard, and sounds a mighty blast.
High are the hills, that great voice reaches far-
They hear it echo full thirty leagues around.
....
Rollant ad mis l'olifan a sa buche,
Enpeint le ben, par grant vertut le sunet.
Halt sunt li pui e la voiz est mult lunge,
Granz .XXX liwes l'oïrent il respundre.[1753-1756]

The acute, terrifying sound is heard echoing loudly through a long expanse of
land, a strong metaphor for the cultural power of the *ostensive*. This great distance
encompasses dramatically the large measure of terrain the departing Franks must
cover doubling back to Roncevaux. The need for an attempt to arrive on time is
not an issue to be debated. Yet actual deployment of the Army as an answer to the
call is questioned by Ganelon [1770-1784]. The first and third Oliphant calls
enclose the insolent speech in which Ganelon impudently attempts to turn the
blowing of the horn into a frivolous gesture, "Just for a rabbit he'll blow his horn
all day!"[1780] [24] Ganelon intends to invert Roland's sacralizing act; the
unabashed insolence intends to turn blowing of the horn into a sacrilege. The
audience senses that retribution hangs in the balance. The quintessential patriot
stands as sharp contrast to the obdurate traitor. Before Ganelon's guilt engages
the reader's attention at length, however, again we hear the horn blown a second
time by Roland.

And now Count Roland, in anguish and in pain,
With all his strength sounds the great horn again.
Bright drops of blood are springing from his mouth,
Veins in his forehead are cracking with the strain.
....
Li quens Rollant, par peine e par ahans,
Par grant dulor sunet sun olifan.
Par mi la buche en salt fors li cler sancs.
De sun cervel le temple en est rumpant.[1761-1764]

_____

[24] *"Pur un sul levre vait tute jur cornant."*[1780]

The sound of the Oliphant, whose reach is vast, requires a painful, even fatal, effort by the dying warrior. Ganelon's assertion that Roland's lavish frivolity, "great pride" [1773],[25] is manifested in the Oliphant's call is clear indication of his guilt in the treacherous ambush. He urges the Army on back to France, questioning the need to stop for a boastful braggart, "Now he is playing some game to please his peers" [1781].[26] But the call does not project a light tone. The response must be a direct consequence of its urgency. The *jongleur* explains the fatal result, as for a third time the Oliphant is blown from Roland's bleeding mouth:

> Count Roland's mouth is crimson with his blood,
> His temples broken by the tremendous strain.
> He sounds the horn in anguish and in pain.
> ....
> *Li quens Rollant ad la buche sanglente.*
> *De sun cervel rumput en est li temples.*
> *L'olifan sunet a dulor e a peine.*[1785-1788]

Oliver's original plea equates the Oliphant's call to the urgent need for timely help. Near death, Oliver's potential *imperative* plea to Charlemagne's Army, "Come and save us," cannot be answered in fact, so value of the Oliphant's call as an instrumental act has waned. We even witness how the Oliphant eventually becomes a mere war mace used by Roland to bash in the skull of a treacherous Saracen [2288-2291]. Roland's use of the instrument is not a true distress call, but a nominal *ostensive* sign informing that he is dying and must soon be buried [1750-1752]. That is, the Oliphant's function regresses from the worldly useful verbal *imperative*, "Come back," to the *ostensive* proclamation, "Here we are, martyrs!" This contrast adheres to the theme of heroism central to the overall epic narrative of the *Chanson*. The main thrust of Roland's being is to attain immortal

---

[25] *"Asez savez le grant orgoill Rollant."*[1773]
[26] *"Devant ses pers vait it ore gabant."*[1781]

glory, *los*, through a heroic death. He does not wish to put down his sword and sound the call for aid.

To announce Roland's extreme bravery in battle is the Oliphant's call true ostensive value. In this manner, by pointing to a sacred scene of sacrifice, the call becomes a sign of perennial patriotism. Such is the actual sense of the famous line that distinguishes between the two peers: *Rollant est proz e Oliver est sage;* "Roland's a hero, and Oliver is wise" [1093]. Untrue to the Oliphant's principal, practical use, Roland dies. But, while the Oliphant's value as an instrument to produce a useful signal diminishes, epic tragedy increases. This transition within a "neoclassical esthetic," while problematizing the scene of tragic representation as a crucial *locus* within the work, encloses a self-contained diachronic analysis for the sequential emergence of the elementary speech forms. Delay in sounding the Oliphant crystallizes appropriate progression in the actual development of language. The *ostensive* sign must first establish significant nominal meaning before a verbal *imperative* sign can perform its useful semantic function, even if in the worldly context reaffirming this priority spells doom in the character's immediate experience.

Through meticulous analysis of the heroic motivation for deferring to sound the Oliphant, we may probe into what the sound means for plot development within the narrative. An emergency call from a great, courageous warrior could only mean that death approaches, hence his allies must exercise revenge against his foes. At Roncevaux the tragic hero reveals his true nature as model warrior in quest of glory. Gans explains: "The premature return of Charlemagne's army would have saved him at the price of abandoning the closure of the scenic locus to which his election had led him and on which he finds martyrdom, the source of ultimate significance." [27] Enhanced by sheer distance from the listener, the call of an Oliphant sounds brutally clear and acquires instant discursive status. As non verbal sign the Oliphant's sound has multiple plausible meanings in the poem:

---

[27] *Originary Thinking* 152.

"Enemies," and, consequently, the unavoidable cause for their presence must be "Treason," and, hence, the conclusion becomes "Death." [28] This triple *ostensive* sign strikes an intensely tragic chord in the listener's mind. In the encompassing and climactic Oliphant's call we sense the condensed narrative of *The Song of Roland*, as the story of the rearguard's ambush, together with its cause and aftermath. Perceiving the striking sound of the horn, even in a silent reading of the poem, we sense the need to immediately grant the sacrificial victim sacred status as martyred hero.

Since the Oliphant is a wind instrument, our view of the central climax of the *Chanson* could in principle be reconstructed strictly from the acute psychological impact caused by its sound. The Oliphant's call is never described in the text; however, overwhelming sound is implied. In his *La Chanson de Geste, essai sur l'art epique des jongleurs* (1955), Jean Rychner explains the dramatic virtue of similar strophes by suggesting that repetition stops the action's narrative flow in order to create a lyrical effect on the audience. The three instances when Oliver asks Roland to sound the Oliphant are matched repeatedly by the latter's obstinate refusal.

Oliver 1051- 1052	Oliver 1059-1060	Oliver 1070-1972
Roland 1053- 1054	Roland 1062- 1063	Roland 1073-1074

The same lyrical effect resounds with tragic clarity during actual sounding of the horn. Oliver explains that clamoring for the need of proper burial is now the only material objective for Roland to sound the Oliphant [1742-1751].

Roland	1754	1762	1787

---

[28] Ganelon's betrayal during his Embassy at the enemy camp leads to the rearguard's separation from the Frankish host and makes possible ambush at Roncevaux by the emir's forces.

After line 1762 the two following verses describe the previously mentioned fatal consequence: blood gushes forth from the mouth of our hero with nearly bursting temples [1763-1764]. The third blowing of the Oliphant is performed by a dying warrior, bleeding from the mouth, who uses his last breadth to proclaim his death before his temples burst from the effort [1786]; Roland thereby invests his last bit of energy into production of a sign proclaiming permanent sacrificial significance.

The rising climax occasioned by the triple Oliphant's call is represented by Charlemagne's reactions as listener. First he recognizes the eminence of raging battle, "Our Franks are in a fight" [1758].[29] Followed by: "That is Count Roland's horn!"[1768] [30] Awareness is crowned by final recognition of the tragic effort: "How long that horn resounds!"[1789][31] All three perceptive responses are introduced by the *jongleur* with the phrase, "then says the king" [1758, 1768, 1789]. Rychner remarks that the French Army's response to the crisis at Roncevaux is acute, immediate, and also arranged by the singer of tales in triple series:

Each to the other pronounces the same vow.

There is not one who can hold back his tears.

Not one but grieves and bitterly laments.[32]

....

N'i ad celoi a l'altre ne parolt.[1803]

N'i ad celoi ki durement ne plurt.[1814]

N'i ad celoi n'i plurt e se dement.[1836]

---

[29] *Ço dist li reis: "Bataille funt nostre hume!"*[1758]
[30] *Ce dist li reis: "Jo oï le corn Rollant!"*[1768]
[31] *Ço dist li reis: "Cel corn ad lunge aleine!"*]1789]
[32] Syntactic repetition is seen more obviously in the original text, since identical understatement heads all three lines. The phrase in question is, "*N'i ad celoi*," literally, "There is not one." Terry 70.

These lines epitomize loyalty to the king. Once each warrior becomes a lethal "Roland," access to the privileged threshold for attaining a glorious death in battle is lowered from the nobility on down to the common foot soldier in order to allow every warrior communal participation in the sacrificial act. The demand for *pathos* provides motivation that propels the dramatic action. Roland is no longer an elite warrior lord, nephew to the king, but an *exemplum* for all to follow. Consequently, the heroic sentiments of our poly-faceted central character become emotionally accessible to the audience. We assume that, once the speaker extends his role and significance beyond speech, and into action, Roland's desire for glory, if not glory itself, is more easily experienced by the audience.

The final feeble blowing of the horn by the moribund hero is intimately tied to the coming of Charlemagne and his forces:

> His temples broken from sounding his great horn,
> Longing to know if Charles is on his way,
> Weakly, once more, he blows the Oliphant.
> ....
> *Rumput est li temples, por ço que il cornat.*
> *Mais saveir volt se Charles i vendrat:*
> *Trait l'olifan, fieblement le sunat.* [2102-2105]

This fourth and last blowing of the horn is pathetic and tragic. The dramatic impact cannot be overestimated. The next mention of the Oliphant in the text is as an object which Roland places under his body, together with *Durendal*, while he collapses, bowing to his promise to die a conqueror, facing the enemy [2359-2360, 2363]. Rychner ends his structural analysis indicating that the effect on the audience of syntactic repetition is strongly musical, paralleling perhaps a nostalgic rhapsody.[33] Viewing the hero's death as sacrifice, we may consider the musical accompaniment of Gregorian chant, integral component for the medieval ritual of

---

[33] Rychner 93-98.

the Catholic Mass, as similar in effect to the lyrical tone Rychner suggests as backdrop for the Roland epic.[34]

We may add to Rychner's explanation the insight that line 1755, "High are the hills, that great voice reaches far," starts the mentioned sequence in *laisse* CXXXIII;[35] a parallel line, "High are the hills, and shadowy and vast" [1830], ends the same sequence of six *laisses* [CXXXIII-CXXXVIII].[36] These two lines are reminiscent of the line, earlier in the poem, outlining the separation of the rearguard, "High are the hills, deep valleys shun the light" [814].[37] The rearguard is left forlorn in a plain shadowed by high cliffs. We remark how the same phrase appears, as variation on a theme, in the *laisse* where Roland exercises his last living prowess as he smashes the Oliphant on the head of the Saracen who had feigned death to surreptitiously capture *Durendal*, even after the rest of Marsile's forces had already fled the field of battle: "High are the hills and very high the trees" [2271].[38] The resounding line cements his tragic stance; from first to last the hero perishes as martyred warrior with his rearguard at Roncevaux, the sacrificial scene.

The Oliphant's call impacts the present and future in the epic plot, creating verisimilitude. Without the scene of betrayal at the enemy camp, the ambush at Roncevaux would not take place. As previously mentioned, the Oliphant's call, a triple *ostensive*, links the Saracen *Enemies* to Ganelon's prior *Treason* and Roland's subsequent *Death*. We sense the surging importance of the Oliphant's sound throughout the narrative. Time and again the characters are aroused into action by memory of the horn's sound. We recall how the trauma itself was succinctly summarized and tied to the narrative context of the *Chanson* by the noble old counselor Naimon at the time the horn is blown. Upon hearing the

---

[34] Forney and Machlis 82.
[35] *Halt sunt li pui e la voiz est mult lunge.*[1755]
[36] *Halt sunt li pui e tenebrus e grant.*[1830]
[37] *Halt sunt li pui e li val tenebrus.*[814]
[38] *Halt sunt li pui e mult halt les arbres.*[2271]

sound, the king's peer extends interpretation, and draws vital meaning for the Oliphant's progressively debilitating call:

> Duke Naimon answers, "Great valor swells the sound!
> Roland is fighting: he must have been betrayed -
> And by that man who tells you to hang back."
> ....
> *Respont dux Neimes: "Baron i fait la peine!*
> *Bataille i ad, par le men escïentre.*
> *Cil l'at traït ki vos en roevet feindre."*[1790-1792]

The call proclaims the surging Saracen attack that annihilates the rearguard. No longer a vague possibility, result of the unmistakable threat rings clear: "We die." Henceforth Charlemagne knows how to view Ganelon:

> "Here is a felon I'm leaving in your charge-
> He has betrayed the vassals of my house."
> ....
> *"Ben le me guarde, si cume tel felon!*
> *De ma maisnee ad faite traïsun."*[1819-1820]

Roland's courage looms large over Ganelon's treason.[39] Thus, the *ostensive* use of the Oliphant affects deeply character motivation and overall plot development by showing Charles that Ganelon's description of the scene is false. The delay in blowing the horn, an attitude of seemingly incomprehensible heroism, proclaims

---

[39] The resentful relationship between Roland and Ganelon has a tradition extrinsic to the text of the *Chanson*. A Latin text from the 1100's, "Song of Ganelon's Betrayal," *Carmen de prodicione Guenonis*, narrates how Charlemagne, tired of fighting, wished to return to France; but Roland insists on sending an Embassy to Marsile in Saragoce demanding surrender before returning to France. Roland suggests that Ganelon should take the king's message to the Saracen emir. There follows a confrontation between Roland and Ganelon, where Charles intervenes. Ganelon goes to Marsile, who becomes enraged at the demand for surrender. Marsile's wife, Bramimonde, calms down the emir. At this point Marsile insinuates the ambush plot to which the frightened Ganelon agrees, later bringing back the lying promise of Saracen surrender to Charles. The king departs, leaving the rearguard behind, a decision which makes the ambush at Roncevaux possible. In the Latin poem, delaying warfare through Ganelon's resentment interrupts the train of narrative, as in *The Song of Roland*. The scene of betrayal at the enemy camp provides a backdrop to events leading up to tragedy at Roncevaux. Menéndez Pidal 129-130.

that Roland cannot be seeking help. Ganelon reads this as mere frivolity, whereas Charles understands the *ostensive* use of the horn as a sign of martyrdom.

After placing Ganelon under arrest, and accomplishing a successful pursuit of Marsile's fleeing forces, Charles and his men double back to Roncevaux. While the French tread among the corpses left at Roncevaux, the king tears his hair and the Army weeps [2412-2415]. As twenty thousand men faint, absorbing vicariously the death of Roland's men, Naimon exhorts the king's forces to avenge the loss and relieve the sorrow: "Let us avenge our grief!"[2428][40] A clear danger is that grief may impair the deployed Army's ability to fight effectively.

In the *Chanson* the hero's sacrifice is an asymmetric gesture since no act by the avenging Army is enough to bring back Roland, nor can Charlemagne's prowess match the obstinate courage of his dead nephew. Due to Roland's *ostensive* gesture, the epic poem cannot become the "Song of Charlemagne," regardless of the hero's absence in the last third of the narrative.[41] Roland feeds the plot as sacrificial victim and therefore determines and dominates the scenic focus of the poem. The glory Roland reaches in death becomes an object sought by the Army of Charlemagne through progressive reiteration of such warring spirit as is kindled by remembrance of the Oliphant's call. After Roncevaux, there follows directly in the text the chase and demise by drowning of Marsile's fleeting forces. The king prays that the sun should not set before revenge is carried out [2450-2451].[42] The consequence is a *deus ex machina*: an angel comes and God performs a miracle [2454-2459]. The sun finally sets after all pagans drown

---

[40] *"Car chevalchez! Vengez ceste dulor!"* [2428]

[41] Moignet quotes P. le Gentil: "Charlemagne avenges Roland. Therefore, the latter remains the character around whom the work holds together; yet, pursuing vengeance, the former remains at the height of his holy mission. Thus, the homage Charlemagne grants Roland ennobles the one, without diminishing the other. There is no *Song of Charlemagne* grafted on a *Song of Roland*; there is a *Song of Roland*, of powerful import, which surpasses the plot to attain the level of myth." Moignet 249.

[42] This scene is reminiscent of Joshua 10:12-14. The Lord delivers up the Amorites before the children of Israel by delaying sundown for a whole day. The delay of nightfall allows the chosen of God to have revenge over their enemies. Scofield 279.

[2474]. Roland reaches *Eternal Glory* through a warrior's death while a guardian angel facilitates victory for the French Army.

We see the Oliphant's sound as intensely climactic due to precise developments in the plot. The unfulfilled desire to help, no longer possible in response to the Oliphant's sound, is redoubled and infuses new strength into the French forces. The French host cannot save the rearguard, but every soldier in the Army can run to battle willing to die valiantly on the rearguards's example and attain *Eternal Glory*. The protracted Oliphant's call spreads the deep significance of the quest for glory throughout the Frankish host. Every soldier is desperate to participate in the violence of war.

The unraveling sacrificial crisis defers the expectation for a victorious end to war and the narrative closure the reader may long for. This is war, not just a single battle. In terms of motivation we sense that to die while fighting brings immortal glory through noble death. As Turpin explained during the thick of the fray at Roncevaux: "Better to die with honor on this field!"[1518] [43] Heaven waits: "For you stand open the gates of Paradise" [1522].[44] The reader senses in the *Chanson* a quest to transcend military and political renown in order to attain a self-justified system of sacred values. At Roncevaux Turpin repeatedly blesses warriors [1034,1515]. He draws his sword, *Almace*, and causes havoc [2089].[45] Throughout, the bishop fights valiantly, and he eventually dies with arms outstretched, emulating crucifixion. [2241]

Before concluding our argument, we should consider briefly the sacrificial nature of Carolingian conquest. Just as Roland wishes to keep *Durendal* from enemy hands, Roland's body is protected by Charles' avenging forces [2434-2439]. The French war-cry *Montjoie*, transformed on Roland's example into an *ostensive* sign, is intimately tied to the concept of deriving victory from suffering,

---

[43] *"Asez est mielz que moerium combatant."*[1518]
[44] *"Seint pareïs vos est abandunant."*[1522]
[45] Bédier considers "Almace" an unexplained name, *nom inexpliqué*, for Turpin's sword. Bédier 506. The etymology of the name is uncertain; but it may be derived from the Germanic, *all macht*, meaning Almighty. *The Free Encyclopedia.* en.wikipedia.org

as seen in the morphological reference to Charles' sword, *Joyeuse*. The mount, or hill, of joy recalls the place where the martyrdom of Saint Dennis took place. Historically, the battle-cry *Montjoie* itself evolved from reference to "Mount Joy of St. Denis." [46] Both swords, Joyeuse and Durendal, bear relics in the pommel establishing the representational value of sacrificial martyrdom.[47] Both swords now have a new sacrificial narrative to carry them beyond their physical status as weapons. A fearless desire to undergo sacred sacrifice in order to attain *Eternal Glory* takes central stage. To keep faith, the French Army had to protect Roland's corpse from violation by Saracen forces upon arrival at Roncevaux. As the body of a saint, Roland's corpse became object of a communal interdiction. Ancestral custom and natural law demand Christian burial. Roland deserves public reverence, for he endured martyrdom. Burial rites must be performed to answer the societal need for propriety [2962-2973]. The individual's dependence on the community for proper signification is absolute. Glory is realized through martyrdom, a relic left to posterity.

At five instances during the course of the narrative the battle cry "Montjoie," *Munjoie*, is voiced by characters in direct discourse. The *jongleur* quotes Turpin directly as the bishop-warrior urges the troops at Roncevaux [1350]; and soon thereafter in the text the Army shouts their war-cry during the furious infighting [1525]; finally, Oliver shouts out one last time as he dies [1974]. The increasing sense of doom strikes a tragic note as the *ostensive* Montjoie

---

[46] From the *Annuaire des dictionnaires* website we read: "The Mount Saint Dennis Joy, or simply Joy Mount, used to be the name for the hill close to Paris where Saint Dennis endured martyrdom; so named because the place of martyrdom was a site of joy for the saint who received his reward. The Mount Saint Dennis Joy, *la Mont-joie Saint Denis*, means the Mount Joy of Saint Dennis, *le Mont-joie de Saint Denis*, according to the old rule which rendered the Latin genitive by the ablative case. The name Joy Mount extended to all mounds and was used even figuratively. On the other hand, the French seized as war cry Mount Joy of Saint Dennis, or, simply, Mount Joy (Montjoie); eventually this war cry became the name for the King's coat of arms in France."(trans. mine) *Dictionnaire de l'Académie Française.* Mediadico.com

[47] *Durendal* has St. Peter's tooth, part of a garment from the Virgin Mary, drops of St Basil's blood, St. Dennis' hair [2344-2348]. *Joyeuse* has in the hilt part of the spear that wounded Christ; as direct referent to death and resurrection, it can conquer any foe [2503-2505, 2510]. Geoffrey of Anjou carries the oriflamme, which belonged to St. Peter, also called *Montjoie*.[3092]

becomes a harbinger of death. The other two instances in the *Chanson* for the protracted battle-cry to appear in direct discourse occur as Charlemagne's Army approaches Roncevaux [2510]; and we again hear "Montjoie" voiced while the host witnesses the need to confront the forces of Baligant, the powerful emir from abroad who comes to avenge the dead Marsile.[3092]

The deep significance of the Oliphant again surfaces as the French contemplate Charles' encounter with Baligant:

> The clear-voiced trumpets ring out from every side-
> Above the others resounds Count Roland's horn;
> Then all the Frenchmen remember him and mourn.
> ....
> *Sunent cil greisle e derere e devant;*
> *Sur tuz les altres bundist li olifant.*
> *Plurent Franceis pur pitet de Rollant.* [3118-3120]

Roland had died in line 2397. But, after the battle at Roncevaux, Roland's memory has become significant enough to maintain continual yearning for warlike action. His Oliphant is a pervasive, phantasmagorical presence. Every trumpet sound harkens back to Roland's awe-inspiring horn. The Oliphant's call must retain its nature as supplemental war-cry despite the intense sorrow. The practical consequence of lamentation over Roland's death must include an expansion of violent warfare. The king battles Baligant at the call of his own Oliphant, as Baligant himself tells his son Malpramis:

> "A valiant lord now sounds the Oliphant,
> From his companion a trumpet call comes back."
> ....
> *"Cil est mult proz ki sunet l'olifant:*
> *D'un graisle cler racatet ses cumpaignz."* [3193-3194]

A trumpet answers in support of the Oliphant's call, enhancing the exchange. Throughout the *Chanson*, cause for sounding an Oliphant must broadcast the urgent desire to kill. The Oliphant no longer prompts the need to signal a new martyrdom; that meaning is already deeply infused within its call as a rally to the Army. The post-Roncevaux Oliphant call, intensified through Roland's death, has become a much more dreadful threat. The enemy is aware of such intensity, as well. Henceforth in the text, and through the ages, the French Army becomes ferociously courageous in their charge against the enemy. In fact, the phrase "high they are," *haltes sunt*, mentioned during discussion of the high pass leading to the Roncevaux valley, occurs again in the final mention of the king's Oliphant, when last blown in the *Chanson*:

> The trumpets sound, their voices clear and high;
> The Oliphant rings out above them all.
> …….. …………………………………………
> *Sunent cez greisles, les voiz en sunt mult cleres;*
> *De l'olifan haltes sunt les menees.* [3309-3310]

By breaking through the tonal resonance of trumpets, an Oliphant's call combines the audiovisual remembrance of sound and scene through dramatic projection of Roland's heroic death, coalescing military prowess into legend The sound of an Oliphant encourages the forces of Charlemagne, as he prepares to encounter his nemesis, Baligant. In context, the phrase, "high they are," *haltes sunt*, referring to an Oliphant's sound, recalls the rearguard's fate at the sacrificial scene, since previously in the poem "high they are", *halt sunt*, had referred to the high cliffs at Roncevaux. The ambush site not only inspires awe, the locale also provides a setting for carrying the realistic echo of an Oliphant's sound, and maintains our attention on the call's climactic significance in overall character motivation and plot development.

At the end of the Chanson, after Ganelon's trial and execution, the Angel Gabriel comes to Charles to signal new troubles ahead [3994-3998]. As the Wheel of Fortune spins on, the king confesses his sorrow: "'God!' says the king, 'how weary is my life'" [4000].[48] Since Charles takes on a more active role after

---

[48] *"Deus", dist li reis, "si penuse est ma vie!"* [4000]

Roland's death, we must refer the Carolingian offensive to the Christian doctrine of redemption through sacrifice. Reward is other worldly and in human life joy is temporary. In the context of Christian piety victory is delayed until the afterlife. The glory Roland attains is fame in this world, salvation in Heaven.

We recall that in his final moments at Roncevaux, Roland uses the Oliphant as physical object to smash the head of a Saracen [2295]. We assume the horn as instrument survives the impact, since the object resurfaces twice in the later narrative [3119, 3310]. But possibly Charlemagne may have his own Oliphant, separate from Roland's horn. The text is not specific in this regard. What is certain is that the effect of the Oliphant's call endures beyond the instrument's use. Its influence cannot be destroyed. We know the hero's sword is indestructible; *Durendal* survives as an emblem of destruction [2342]. Once Roland dies heroically, his soul is carried to Paradise by a Cherubim, and by both, Saint Michael, and Saint Gabriel [2393-2396]. He will not wield *Durendal* again, and the Oliphant will be blown by him no more. But his glory and heroism survive his demise. Since through self sacrifice a Christian may strive toward *Eternal Glory*, we sense a movement from Christ to Christianity, or from Roland to the Crusades of ensuing centuries. At Roncevaux a center with an absent central figure opens up potentially infinite plural signification.

Careful reading of *The Song of Roland* discloses a sacrificial crisis caused by rivalry, and persistent, unmitigated violence, intensified by the Oliphant beyond literary language. The dramatic impact of the hero's sacrificial death acquires greater symbolism through the delayed Oliphant's call. Roland's delay in blowing the horn, converting an *imperative* signal into an *ostensive* sign through authentically sanctifying heroism, expands epic drama and suspense immeasurably, while conveying the real spirit behind a genuine call to arms from the distant past.

Works Cited

Bédier, Joseph. *La Chanson de Roland.* Vol 2. *Commentaires.* L'edition d'Art H. Piazza, 1968.

Dictionnaire de l'Académie Française. 8 ème édition. 1987-2010. http://www.mediadico.com/dictionnaire/definition/mont-joie/1

Farnsworth, William Oliver. *Uncle and Nephew in the Old French Chansons de Geste, A Study in the Survival of Matriarchy.* Columbia U P, 1913.

Forney, Kristine, and Machlis, Joseph. *The Enjoyment of Music.* W.W. Norton & Company. 2008

Gans, Eric. *Originary Thinking: Elements of Generative Anthropology.* Stanford U P, 1993.

------------ *The Origin of Language. A New Edition.* Spuyten Duyvil, 2019.

Guillaume d'Orange: Four Twelfth-Century Epics. Translated by Joan M. Ferrante. Columbia U P, 2001.

Jotischky, Andrew, and Hall, Caroline. *Historical Atlas of the Medieval World.* Penguin Books, 2005.

La Chanson de Guillaume. Edited and Translated by Philip E. Bennett. Grant & Cutler LTD, 2000.

La Chanson de Roland. Edited by Gérard Moignet. Bibiothèque Bordas, 1969.

Le Charroit de Nîmes. Translated by Fabienne Gégou. Editions Champicn, 1984.

Menéndez Pidal, Ramón. Translated by I.M. Cluzel. *La Chanson de Roland et la tradition épique des Francs.* Editions A. et J. Picard, 1960.

Rychner, Jean. *La chanson de geste, essai sur l'art epique des jongleurs.* Genève: Librairie E. Drcz, 1955.

The New Scofield Study Bible: New King James Version. Edited by C.I. Scofield. 1967. Consulting editor Arthur L. Farstad. Thomas Nelson Publisher, 1989.

The Free Encyclopedia. http://en.wikipedia.org/wiki/Almace

The Song of Roland. Translated by Patricia Terry. Bobbs-Merrill, 1979.

Shared Guilt for the Ambush at Roncevaux

In the *Roland* epic we contemplate a prominent but compact framework within a vast landscape, a hut on a hill. The overall struggle between Christians and Saracens in *The Song of Roland* gives way to the attention we must bestow on the friction arising among heroes in the French camp, Charlemagne, Ganelon, and Roland. Within the poem the Christian-Saracen struggle is not explained in terms of worldly causality; several times, rather, one side is identified as good, and the other as evil.[1] There is a basic rivalry which is not the crucial dissension of the epic. As in *The Iliad,* where Achilles' rage is directed, not toward the Trojans, but against Agamemnon, the rivalry that drives the narrative in *The Song of Roland* is within the French camp. Guilt for the ambush at Roncevaux is shared by the three main protagonists. Ganelon is married to the king's widowed sister, Roland's mother. Before departing on the fateful Embassy to the enemy camp, Ganelon tells Charles: "Remember this: your sister is my wife" [312].[2] Although she is not mentioned by name in the text of the Chanson, as an absent figure of desire, her image looms behind the surging resentment between characters which gradually leads to conflict. The big fight breaks away into internal turmoil.

In our analysis we address first the dreams of Charlemagne, which reveal the pathetic plight of the king in terms of the choices he makes and those he fails to make. We then shall expose the poetic framework in terms of a quadrangular family relationship central to plot motivation in *The Song of Roland.* Both, the monarch's dreams and the family relations, combine in the narrative of the poem to drive Ganelon into the role of scapegoat at the end.

---

[1] "Pagans are wrong, the Christian cause is right." *Paien unt tort e chrestiens unt dreit* [1015]. And again: "'Their cause is evil, and we are in the right.'" *'Nos avum dreit, mais cist glutun unt tort'* [1212]. Translations into English are from the Patricia Terry translation, which I have modified slightly, in order to suit context, twice, (cf. lines 735, 2567). Terry 41, 48. Quotes in the original French are from the Gérard Moignet edition. Moignet 92, 104.

[2] *Ensurquetut si ai jo vostre soer.* [312]

Charlemagne's Dreams

The dreams play an important role in the plot. We find textual evidence of antagonism between Ganelon and Roland, yet the actual reason for Ganelon's betrayal in the *Chanson* is a difficult issue to address directly. Enmity toward Roland definitely translates into a breach of loyalty against the king. On the other hand, the dreams of Charlemagne show a glaring premonition in the text upon which the Emperor refuses to act. Roland himself is too bent on destruction of the enemy to cherish the idea of returning to France. The discord between the three main characters of the *Chanson* remains an issue to be resolved. We contend that their mutual enmity condemns the rearguard to destruction.

The dreams occur in two pairs, before the separation of the rearguard, and after the French host's revenge against Marsile. The crucial decisions: a) to separate from the vanguard, turned rearguard, and b) for the host to turn around and head for France, both shall prove fatal. Since neither decision had been made before the first dream sequence, the content of the dreams reveal their premonitory quality. Aware of the king's failure to act at a crucial moment in the plot, Erich Auerbach ascribes to Charles "somnambulistic paralysis."[3] In the Emperor's first dream of the early sequence, the Count who named Roland to the guard is identified as the king's attacker, who breaks the spear in his grasp:

> Clasped in his hands he holds the ash wood spear:
> Count Ganelon wrenches it from his grasp,
> With raging strength shatters and breaks the wood,
> And sends the splinters flying against the sky.
> ....
> *Entre ses poinz teneit sa hanste fraisnine;*
> *Guenes li quens l'ad sur lui saisie;*
> *Par tel aïr l'at estrussee e brandie*
> *Qu'envers le cel en volent les escicles.*[720-723]

---

[3] Auerbach 101.

Evidently, Ganelon exerts a virulent hold on Charles. The imagery of the forceful grab which sends splinters flying is unforgettable.[4] The physical attack is a clear affront to the king's safety and supreme authority. The crucial issue is the open identity of Ganelon as aggressor.

In the second dream of the first sequence Charlemagne dreams he is at home, where a beast bites his right arm.[5]

> After that dream another vision came:
> He was in France, in his chapel at Aix.
> A vicious beast was biting his right arm.
> Out of the forest he sees a leopard run,
> And he himself it cruelly attacks.
> From his great hill a boarhound rushes out
> And comes to Charles, running with leaps and bounds,
> Seizes the beast, biting off its right ear.
> … …………………………………………
> *Aprés iceste d'tre avisiun sunjat,*
> *Qu'il ert en France, a sa capele ad Ais.*
> *El destre braz li morst uns vers si mals;*
> *Devers Ardene vit venir uns leuparz,*
> *Sur cors demenie mult fierement asalt.*
> *D'enz de sale uns veltres avalat.*
> *Qui vint a Carles lé galops e les salz;*
> *La destre oreille al premer ver trenchat.* [725-732]

The forested Ardenne region is the land of Ganelon's relatives [2558]; but the dark woods also harbor the dreadful mystery of the unknown. [6] Together, the

---

[4] Flying splinters from lances held by knights is a repeated motif during violent jousting in *The Nibelungenlied.* Edwards 121, 125, 170.

[5] The bear who bites Charlemagne's right arm is identified as Ganelon due to the furs he wears. He rises at the Assembly, "Casting aside his cloak of marten furs." *De sun col getet ses grandes pels de martre* [281]. First mention of the right arm metaphor occurs at the Embassy in the Saracen camp, where Ganelon proclaims that, through attack, the French could lose "the Emperor's right arm." *Dunc perdreit Carles le destre braz del cors* [597]. In this same speech at the enemy camp Ganelon equates the loss to future stability for both sides, since the Army would then return to France leaving Marsile to reign calmly over Saragossa: " All of his Empire would be restored to peace.'" *'Tere Major remendreit en repos.'* [600] Thus, the treason seems erroneously motivated by a quest for peace. Such naïve ignorance adds charm to Ganelon's role as villain.

[6] "From Ardenne, *Devers Ardenne,* brings … a specific sense: in poetic terms because Ardenne is, in the poetry of the Middle Ages, the forest of marvels; and in terms of geography because

contrast evokes the vicinity of home mingled with a sensation of dread. We tend to identify domesticated animals with the *good*, and exotic beasts with an enemy.[7] Despite the apparent premonitory nature of this vision, Charles remains impervious to danger, and the ambush at Roncevaux unfolds in the narrative. The bear attacks the king's right arm, *destre braz*, and a leopard's lunge follows. According to Joseph Bédier: "the bear who bites Charlemagne's arm is Ganelon, the leopard is Pinabel, who shall affront the king as Ganelon's champion, the hound is Thierry, who shall face Pinabel on Charlemagne's behalf."[8] The bear is definitely Ganelon, but the leopard could also represent Marsile; and the vicious dog is perhaps a figure of Roland, who heads a smaller contingent.[9] Joseph Duggan remarks: "Critics have been divided over the meaning of Charlemagne's second dream, which some see as foreshadowing Roncevaux, others the Trial of Ganelon."[10]

The dreams may be taken to represent: 1) Ganelon's betrayal, 2) the battle at Roncevaux, 3) the Baligant encounter, and 4) the trial at the end, respectively.[11] Such symbolism seems sensible enough, except that the second dream also encloses possible reference to the trial by combat.[12] Although the reference to

---

it extended up to Vesdre, three or four leagues to the south of Aix-la Chapelle." Bédier 108. (trans. mine).

[7] Benjamin M. Semple says: "While they (scholars) debate over the figures represented by the leopard – is it Marsile? Pinabel? – or over the figures symbolized by the greyhound – is it Roland? Thierry? – they instinctively sense that the greyhound is a *good* figure and the leopard is a *bad* one. I know of no scholar who has suggested that the leopard is Roland or Thierry, or that the greyhound is Ganelon or Marsile." Semple 29.

[8] Bédier 107. (trans. mine).

[9] C.M. Bowra mentions how the beast and the leopard may represent Marsile and Baligant, adding: "there is no need to be too precise about this." Heroic 296. Excessive precision could be misleading. W.G. Emden comments on the view Karl-Josef Steinmeyer presents in his book on the dreams of Charlemagne. Steinmeyer 40-41. "Steinmeyer makes much of the argument that, while Ganelon's trial is foretold twice according to the traditional view, the battle of Roncevaux receives no mention if laisse LVII is held to be the trial." Van Emden 262. T. Atkinson Jenkins, in his edition of the *Chanson*, also explains the second dream as symbolic of the battle at Roncevaux. Jenkins 60.

[10] Duggan 79.

[11] Van Emden 260.

[12] Whitehead 189. Owen 201. Braet 12-13. Herman Braet says: "More often than not, however, both interpretations seem possible to maintain, and the text can be read in two levels." Braet 15. (trans. mine).

Roncevaux is appealing, there is no reason to believe that adopting one interpretation over the other is misleading or contradictory, due to the ambiguous nature of actual dreams.[13] We may consider neither interpretation for the second dream to be mutually exclusive of the other if we focus on the essential gist: the dream may forecast simultaneously an assault against the French forces at Roncevaux and against royal supremacy at Aix. In either case, we should agree that the nightmarish quality of the dreams is adequate to express the unnerving betrayal. The difference in interpretation simply points to the temporal span of the king's premonition regarding different assaults against royal authority in the narrative, early or late in the plot. Such an explanation does not resolve the controversy, but should establish contextual relevance despite the ambiguity.

The second pair of dream visions occurs after the ambush and obliteration of the rearguard; that is, after the king's Army wreaks havoc against the Saracen forces to avenge the loss of the rearguard, and just before confrontation with the forces of Baligant, the emir who brings Saracen reinforcements from abroad. In the king's second sequence of dream visions, we see an expansion of the threat to Charles. Reinforcements from abroad come to avenge Marsile. Baligant's forces become a great number of monstrous enemies attacking Charles' men:

> With great dismay Charles sees his knights attacked
> By vicious beasts – by leopards and by bears,
> Serpents and vipers, dragons and devils too,
> And there are griffons, thirty thousand and more,
> All of them leaping, charging against the Franks.
> ....
> *En grant dulor i viet ses chevalers.*
> *Urs e leuparz les voelent puis manger,*
> *Serpenz e guivres, dragun e averser;*
> *Grifuns i ad, plus de trente millers:*
> *N'en i ad cel a Franceis ne s'agiet.* [2541-2545]

---

[13] Semple 27-29.

Dragons, leopards, vipers, and bears attack the Franks. Threats multiply as a consequence of Ganelon's betrayal.

The king's reciprocal concern over the Army is heightened. Powerless, the sleeping monarch despairs in the nightmare as he regards his men: "The Franks who cry, 'Charlemagne, help us now!'" [2546] [14] This cry for help takes the place of the attack against the king that leaves him unarmed in the first dream of the earlier sequence, for in both cases the monarch is evidently helpless. The inability of the king to run to their aid intensifies the attack against Charles:

> And overwhelmed by pity and by grief,
> He starts out toward them, but something interferes:
> A mighty lion springs at him from a wood,
> Fearful to look at, raging and proud and bold;
> He leaps, attacking the person of the king.
> Grappling each other they wrestle violently:
> But who will rise a victor, who will fall?
> ....
> *Li reis en ad e dulur e pitet;*
> *Aler i volt, mais il ad desturber:*
> *Devers un gualt uns granz leons li vient,*
> *Mult par ert pesmes e orguillus e fiers,*
> *Sun cors meïsmes i asalt e requert*
> *E prenent sei a braz ambesdous por loiter;*
> *Mais ço ne set liquels abat ne quels chiet.*[2547-2553]

The king wants to aid his men but is restrained by a lion. There is a threat against Charlemagne, and the king, unable to delegate his defense to anyone, must withstand the personal attack against himself directly. This blow to the monarch's safety and authority is seen in the uncertain outcome of the ensuing struggle, for Charles' powerlessness reflects the vulnerability of his men in facing Baligant's forces. The expression for an indecisive victory which had provided closure for the second dream of the first sequence is repeated in altered form:

---

[14] *E Franceis crient: 'Carlemagne, aidez!'* [2546]

They don't know which side will win the fight.

. . . .

But who will rise a victor, who will fall?

. . . .

But he could not see which one of them would lose.

Il ne seven; liquels d'els la veintrat. [735]

. . . .

Mais ço ne :et liquels abat ne quels chiet. [2553]

. . . .

Mais ço ne :et liquels veint ne quels nun. [2567]

The adversative conjunction *mais*, "but," introduces a negative hypothesis; this formulaic expression caps both dreams of the second sequence. Noticing the subject of the main verb in line 735, *il* > modern French *ils*, we realize that it was the Army that pondered over the outcome while contemplating the fight. The phase of the attack against royal authority at that time was in the planning stages of a mishandled diplomatic tangle. After the first sequence of dream visions, we have the nomination of Roland triggering the actions heralded by the return of Ganelon from the enemy camp in the scene preceding description of the sleeping Charles. The action is still in an embryonic stage, and the ambush of Roland had not yet taken place. After the second dream sequence, however, Marsile is on the run, bleeding to death due to the injury inflicted by Roland [2574].[15] Through bitter reminiscence of the *right arm* metaphor, the effects of Roland's death are again brought to the foreground, for Charles' liability against Baligant is parallel to the danger his nephew faced in the ambush at Roncevaux. The king subconsciously suspects further trouble.

The second vision of this final dream sequence also sets the stage again, more along the lines of a diplomatic issue than as a strictly military encounter, reflecting the ambiguity we saw in the second dream of the earlier sequence:

Later that night he had another dream:
He was in Aix; on a dais he stood,

_____

[15] "Marsile's right hand was cut completely off." *La destre main ad perdue trestute.* [2574]

107

Holding a bear bound tight with double chains.
Thirty more bears came out of the Ardennes,
Each of them speaking exactly like a man.
They said to Charles, "Sire, give him back to us!
It isn't right for you to keep him here;
We cannot choose but bring our kinsman help."
Out of the palace there came a hunting dog
Who then attacked the largest of the bears;
On the green grass apart from all the rest,
While the king watched, they fought a dreadful fight.
But he could not see which one of them would lose.
....

Aprés icel li vien un'altre avisiun,
Qu'il ert en France, ad Ais, a un perrun,
En dous chaeines si teneit un brohun.
Devers Ardene veeit venir .XXX. urs,
Cascun parolet altresi cume hum.
Diseient li: "Sire, rendez le nus!
Il nen est dreiz que il seit mais od vos;
Nostre parent devum estre a sucurs."
De sun paleis uns veltres i acurt;
Entre les altres asaillit le greignur
Sur l'erbe verte, ultre ses cumpaignuns.
La vit li reis si merveillus estur;
Mais ço ne set liquels veint ne quels nun. [2555-2567]

The second dream in both series of double visions is introduced in similar fashion by a formulaic expression, which sets the sleeping monarch at home [cf. 724-726; 2555-2556]. But the semiotic content of the vision suffers alterations that parallel the events at Ganelon's future trial. For instance, the bear that had bitten Charles' right arm is in chains, and from Ardenne thirty bears come to aid their relative. The benevolence of the king is in question because the thirty bears appear to be in quest of deliverance; but Charles' policy does not gain any benefit from liberal magnanimity. As previously mentioned, the earlier series of double visions had moved from Ganelon's treason in the first dream to suggestions in the second dream of battle at Roncevaux, or Pinabel's challenge at Ganelon's trial, and the subsequent duel. This later dream sequence unveils the king's inability to aid his men, who are attacked by countless foes, and the future assault against royal authority at Ganelon's trial. We go from a very cruel physical reality to the ideological ramifications foreseeable in the future as its result. The king's concern in both visions escalates due to worsening conditions; hence the implications lean toward a marked greater need for prevention. There must be an appropriate reaction to a crisis in order to quell future crises. The hunting dog charges on, engaging the biggest of the bears; and again the outcome seems ambivalent [2567]. The single bear could represent Pinabel, but the focus on the thirty relatives turns our attention more toward the direct parliamentary challenge against the king's authority, rather than to the trial by combat. The duel itself is not as much in the foreground as is the royal authority that is being contested and challenged. A careful reader notices that, while the first sequence of dream visions shows a crisis, which turns from the diplomatic inception of Ganelon's treason to military implementation of the ambush at Roncevaux, the second series progresses in symbolic representation from the suggestion for direct military confrontation on toward reference to the need to arrive at judiciary adjudication; that is to say, from Baligant's confrontation and the Pinabel / Thierry duel we move on toward a need to achieve a lasting resolution for preventing any future legal friction occasioned by ensuing antagonism similar to the obstinacy displayed by the thirty relatives and the barons who favor Ganelon at court [3807-3810]. Differentiation between friend and foe must be clearly established.[16] To end ambivalence in dispensation of justice, Ganelon's guarantors must be chastised by the court [3933]. At the trial Charles rises to the occasion because suffering strengthens the aged figure of the king [3953-3955]. Consequently, the audience is drawn to a more mature Charlemagne.

---

[16] In his *Violence and the Sacred*, René Girard remarks how failure to differentiate leads to lack of social order. Violence 124-125, 146, 245.

Quadrangular Relationship

In her *Matriarchy, Patriarchy, and Imperial Security in Africa*, Marsha R. Robinson explains:

> Many of the Celts, Cantabrians, Picts and Teutons had laws by
> which property, especially land, was inherited from one's mother
> and her brother. If a foreigner were to marry into such a family,
> say a soldier to a local woman, he could not inherit the land. It
> would be controlled by the bride and the bride's male kin. A
> foreign husband was indebted to his brother-in-law who controlled
> the center of wealth accumulation. A foreign soldier could not
> transfer his wife's inheritance to the empire that he served. Wealth
> remained within the bride's family. This is called matrilineal inheritance. [17]

Glorification of the nephew can be seen as one of the main features of a system which establishes matrilineal inheritance. W.O. Farnsworth in his study *Uncle and Nephew in the Old French Chansons de Geste* insists that in a primitive state of civilization matrilineal descent is important in tracing heritage. [18] Farnsworth argues that the matrilineal tracing of descent goes back to an earlier time, preceding the patriarchal trends prevalent in the Roman Empire:

> The introduction of Christianity and of Roman influence among
> the tribes of the north must have been the most important factor
> in the transition to paternal authority. [19]

Nephew-right emerges from the earlier Teutonic tradition of kinship established through the maternal uncle. Farnsworth starts his study by explaining:

> Our modern conception of the family as consisting of father,
> mother, children would at first thought seem to go back in an
> unbroken line to Roman laws, so that it is puzzling to discover
> that French literature of the Middle Ages, in its delineation of
> certain aspects of family life, shows markedly the influence of
> the earliest state of human society about which we have information.

---

[17] Robinson 24.
[18] Farnsworth 1, 157-158, 198, 212, 229, 239.
[19] Farnsworth 243.

As a matter of fact the Old French *Chansons de Geste* show plainly
that there existed in the eleventh, twelfth, and thirteenth centuries,
in the form of tradition at least, a survival of an earlier condition in
which the family was based upon the matriarchal principle. [20]

The term *matriarchal* seems to be a misnomer since the inferior position of the
medieval woman is apparent. [21] Yet the uncle-nephew relationship reveals the
importance of matrilineal descent in the continuation of family tradition. For this
reason Farnsworth considers the prevalence of nephew-right in the *Chansons de
Geste* to have sentimental and not legal roots.[22] Survival of the belief in matrilineal
inheritance is not based on female supremacy as such, but rather on the fact that
in very ancient times the physiological basis for paternity was relatively unknown.
[23] At the advent of civilization, a child just born was very obviously given birth
by a specific woman of the tribe. In some cases, however, it was not possible to
establish definite paternity. Since it was known who the mother was, but not easily
determined who was the father, in order to preserve lineage, property and power
were not transmitted from father to son, but from a man to his sister's son. The
looser the bond between husband and wife, the closer the tie between a wife's
brother to her as sister, and, hence, the closer a relative a maternal uncle could
become to his nephew. The uncle-nephew relationship between Charlemagne and
the hero leads the king to feel extreme responsibility for his safety. Before and
after the ambush at Roncevaux, the attempt to justify his nephew's death torments
Charlemagne.[24]

Roland's predicament affects the pathology of Charlemagne. Since Charles
is reluctant to justify the loss of Roland, he avoids consciously viewing his dream

---

[20] Farnsworth 1.

[21] Farnsworth 243.

[22] Farnsworth 244.

[23] Farnsworth 1.

[24] If the maternal uncle has the cultural importance Claude Lévi-Strauss assigns to the
relationship in his *Structural Anthropology* (32, 39-41, 322), then the patriarchal competition
between Charles and Ganelon could escalate to a higher level. Ganelon's role as stepfather of
the hero becomes subordinate to the position held by Roland's maternal uncle, the King of
France. The monarch may have a much greater responsibility for the safety of his nephew.

visions as indicative of a need to restore order in the realm. Yet when Count Naimon begs for an explanation for the king's downcast demeanor, Charles discloses contents of the vision to his noble liege. A premonition of disaster is already evident to Charles in the first dream of the early sequence, as he soon thereafter discloses to Naimon:

> 'I can't keep silent the sorrow that I feel,
> For Ganelon will be the doom of France.
> Last night an angel sent me a warning dream:
> I held a spear – he broke it in my grasp,
> That count who named my nephew to the guard.'
> ....
> 'Si grant doel ai ne puis muer nel pleigne.
> Par Guenelun serat destruite France.
> Enoit m'avint un avisiun d'angele,
> Que entre mes puinz me depeçout ma hanste;
> Chi ad juget mis nés a rereguarde!' [834-838]

The treason of Ganelon and death of Roland become the anguish of Charlemagne.

Any truce or delay in the fighting is temporary. We recall that the *Chanson* begins with the format "Assembly-Embassy/Assembly-Embassy." The Assembly at the Saracen camp brings out the need for a stratagem to send Charlemagne's Army back to France. The Assembly at the French camp results in a confrontation in which friction between Ganelon and Roland becomes obvious, stepfather to stepson. In her translation of *The Song of Roland* Patricia Terry remarks that in medieval society the *parrastre/fillastre* relationship was viewed unfavorably by both parties. [25] In the context of the Assembly, Roland recalls that previous efforts to reason with Marsile resulted in death for the ambassadors, Basan and Basile. Obeying precedent, Roland reasons as a warrior, "Finish the fight" [210], "To Saragossa lead on" [211], and "Avenge those men" [213]. [26] The language is

---

[25] "Ganelon is indeed Roland's stepfather, but the word used here, *parrastre*,[277] is an insulting one; similarly *fillastre* in line 743." Terry 14.
[26] '*Faites la guerre,*'[210]; '*Metez le sege,*'[211]; '*Si vengez.*'[213]

appropriate for wartime. Roland wishes to advance onward and take over Saragossa. Charlemagne's Army should not turn back. Here is the main point of contention between the monarch and his nephew. Precedent belies the wisdom of sending another embassy to Marsile, or trusting his word in any way. Yet, understandably, Charlemagne's friction with his nephew is sheltered in the subconscious mind. The king does not adjudicate through the wisdom of precedent. As we have seen, only in a dream sequence can Charles face his guilt. After the perturbed monarch awakens from the first series of dream visions, he asks the leaders among his men who shall guard the narrow pass to cover his retreat. Ganelon tells the king to appoint Roland [743]. Subconsciously aware of intense risk, Charlemagne exclaims "Vile demon that you are!"[746]; [27] nevertheless, Roland stays behind. The king's unwitting trust of a seemingly treacherous enemy assigns to him the role of subconscious rival in relation to his own nephew.

Roland's leading role, urging war in the Assembly at the French camp following the Saracen Embassy, is similar to the aggressive Blancandin's role in the enemy camp. Blancandin is ready to sacrifice hostages, even his son, in order to advance a military goal [149]. Roland is always prepared to undertake personal risk in the service of a perennial vow to uphold chivalric honor to the end. On the other hand, Ganelon, with resentment due to injured pride, presents an idiosyncratic distinction in heroic roles which has military implications in a warring society. The resentful Ganelon seizes the opportunity treason affords to become a figure central to plot development during the Embassy at Marsile's camp.

As surrogate father figure to Roland, and as King of France, Charlemagne must safeguard posterity, whatever sacrifice this longing may entail. The readiness to implement filial sacrifice links Ganelon and the Saracens, for they have decided to intentionally send their sons, and a portion of their wealth, as

---

[27] *'Vos estes vifs diables.'* [746]

guarantors for a truce based on a false oath, all in order to get Charlemagne to turn back and leave Roland exposed to treacherous attack [40-45]. Leaving behind the rearguard involves a difficult personal decision for the king; yet assigning duty to Roland as head of the rearguard is a move justifiable in terms of military strategy. Charles ponders over who shall take the vanguard duty once Roland is in the rear guard: "And in the vanguard – who'll have the leader's place?" [748] [28] Since Roland's forces ride point on the war trail, the frontward flank switches roles in order to provide protection for retreat. Charlemagne's column, on the other hand, turns around in formation while the Army simply shifts course, now headed by Ogier of Denmark [749]. The change in direction of the Carolingian host, their *about face*, dramatizes the separation. [29] The irreplaceable young hero remains facing the enemy willingly in order to protect the monarch. Indirectly, and however reluctantly at the conscious level, Charles compromises his nephew's safety.

This intricate family involvement appears to originate in a secret rivalry and revolve around the longings it engenders. An image of sister, wife, and mother emerges as structural principle. The text of *Gui de Bourgogne*, another medieval *chanson de geste*, mentions Dame Gile, identified as duchess and Charles' sister. [30] Farnsworth collects two passages from *Gui de Bourgogne* in Appendix A of his book. [31] Dame Gile is identified twice as sister of Charlemagne, wife of Ganelon, and mother of Roland:

> And the king Gui immediately summoned dame Gile:
> She was the sister of Charlemagne, king of Saint Denis,
> And wife of Ganelon, whose body God damned,

---

[28] '*E ki serat devant mei en l'ansguarde?*'[748] A few lines before, the king had asked who would be at the rearguard, and Ganelon had nominated Roland for the fateful post [742-743]. Charles appoints Roland, "While in his eyes unwilling tears appear." *Ne poet muer que des oilz ne plurt.* [773]

[29] As the old vanguard turns rearguard, with simultaneous selection of a new vanguard, the *jongleur* exclaims sententiously: "You have no baron who will dispute that now." *N'avez baron ki jamais la remut.* [779]

[30] Farnsworth 213.

[31] Farnsworth 244.

And was mother of Roland of the courageous heart.
Et li rois Guis tantost fait mander dame Gile:
Cele ert suer Karlemaine, le roi de Saint Denise,
Et fame Ganelon, qui le cors Dieu maudie,
Et ert mere Rollant à la chiere hardie.[1589-1592]

....

It's Gile the duchess, honored with a gentle heart,
Who is sister of Charlemagne, the strong crowned king,
And wife of Ganelon, the astute peer,
And she is mother of Roland, the renowned knight.

C'est Gile la duchoise, au gent cors onoré,
Qui suer est Karlemaine, le fort roi keroné,
Et fame Ganelon, le compaignon hardré,
Et est mere Rollant, le chevalier menbré.[2920-2923] [32]

Dame Gile's son, and the king's nephew, is appropriately a Count, yet Ganelon, her second husband, is not a Duke, as was Roland's late father, Milon d'Anglers, Duke of Brittany, never mentioned in the *Chanson*.[33] She is not in the story because relationships take precedence over actual names in the *chansons de geste*.[34] We know that the rivalry between stepfather and stepson exists. Charlemagne's failure to act upon his dreams conceals an ambivalent feeling

---

[32] Guessard 49, 89. Translations are mine. In this charming *chanson de geste*, Charles has been away for years, so back in France the sons of veterans away fighting with Charlemagne want a new king at home. They elect Gui de Bourgogne, son of Sanson de Bourgogne, and one of Charles' nephews. Blindly loyal to Charlemagne, and aware of his new role as king, he orders the thousands of men he can muster to go with him to aid Charles. The women complain that Charles took their husbands, now they cannot lose their sons too, so the ladies go along. Gui's sworn duty is to conquer lands for Charles, and to meet up with him at *Luiserne*, where Charlemagne awaits patiently the fall of the city he has had under siege for four years. Meanwhile, Gui conquers *Casaude*. He also conquers *Montorgueil*, on his way to aid Charles. Eventually, after further conquests, Gui meets up with Charles. He offers the king his sword, helps him win over *Luiserne*, and draws as fief for himself Spain. The sons are finally allowed to greet their parents. Bertrand and his father, Duke Naimon, hug and kiss, as do Gui and Sanson. The ladies are overjoyed. Dame Gile comes forward first, followed by Aude [4000-4001]. The epic closes as Charles grants eight days of rest to his troops so they can be with their families. Roland gets together with Aude. Charles insists that all the ladies go back to France as he and his Army, now swollen with provisions and reinforcements, make their way to Roncevaux.
[33] Farnsworth 201, 244. Paris 378.
[34] Farnsworth 163.

toward his nephew which he cannot consciously express. The emperor feels guilt but does nothing. He must support Roland and oppose Ganelon, yet he treats both in a symmetrical fashion. Roland suggests Ganelon go on the Embassy to the enemy camp, and Ganelon nominates Roland for the rearguard; they both suffer death as a result of their actions within the quadrangular relationship. Charlemagne fails to reconcile the two characters. We sense the impossibility for Charles, Ganelon, and Roland to display a proper relationship to each other due to an uncontrollable secret jealousy leading to a hidden subconscious resentment.

The importance of the Duchess, absent from the text, was probably better known to the *jongleur's* medieval audience than it is to the modern reader of the *Chanson*. To recapture the original experience of the tale we resort to extrinsic literary analysis; we dare pose her relevance as a structural principle in *The Song of Roland* to achieve a greater psychological insight into plot motivation. In the mechanics of mimetic desire the subject is torn between two passions: a) love for the object, and b) hate for the rival. Girard remarks about mimetic desire that: "Our first task is to define the rival's position within the system to which he belongs, in relation to both subject and object. The rival desires the same object as the subject, and to assert the primacy of the rival can lead to only one conclusion." [35] Since subjects concentrate on each other, desire becomes displaced; eventually the object of desire is lost in the squabble. Charlemagne loves Roland but does not protect him as he should. The ensuing violence caused by conflicting desire is a shared guilt. In *The Song of Roland* enmity eventually converges against Ganelon, stepfather to Roland and husband of Charles' sister. The unmentioned Dame Gile becomes a place holder in the quadrangular scene of mimetic desire.

Charles must champion political ideals of reciprocal brotherhood. Unquestionably, the personal realm extends into the public arena when commanding an army. The final conversion of Bramimonde, Marsile's wife,

---

[35] Violence 145.

pleases the king, who finally rests peacefully [3989-3992]. The womanizing tendencies of the historic Charlemagne are recorded. Charles had eighteen children with eight of his ten wives or concubines (four wives, six concubines). [36] Farnsworth remarks that a scandalous legend arose after the death of the historic Charlemagne, attributing incestuous intercourse between the king and his sister, before her marriage to Milon, and the birth of Roland soon thereafter. [37] Farnsworth refers to the *Histoire Poétique* by Gaston Paris.[38] In this text the author mentions how in the *Karlamagnus-Saga* Egidius, while celebrating Holy Mass, is said to have had an apparition by Saint Gabriel, who gave the priest a letter; in the letter the priest was ordered to marry the king's sister to Milon d'Anglers. The priest complies and the king confers upon Milon the Duchy of Brittany. The son born seven months after the marriage was believed to have been begotten by Charles.[39] Further on Paris quotes from a XIVth century poem, *Tristan de Nanteuil*, where a sin attributed to Charlemagne is considered too grave to mention. [40] Along the same vein, Farnsworth remarks that the despicable legend eventually became too distant from reality to be true, citing the last page of the Italian chronicle *Li Reali di Francia*: "...it was commonly held that Roland was son of Charles, which was contrary to reality; the king loved him for his virtue and because he saw him courageous of body and soul." [41] Apparently, we may surmise that in the course of history the closeness of Charles to Roland had to be explained beyond the context of nephew-right. Although the paternity issue is probably an exaggeration, the king's closeness to his sister is an aspect of the legend we may preserve without compromise. Robinson proffers an explanation for this exaggeration:

---

[36] The Free Encyclopedia en.wikipedia.org
[37] Farnsworth 213.
[38] Farnsworth 214.
[39] Paris 378.
[40] Paris 381-382.
[41] Gamba 479. The translation is mine.

William Farnsworth pressed for indigenous matriarchy in Western Europe. He did so in his study of the French *Chansons de Geste...* His search was possibly influenced by nineteenth century hegemonic identification with the ancient Roman Empire...The Roman father-son model has been implanted to the point that Charlemagne is recast as the incestuous father of Roland, whose mother was either Gile or Bert, both sisters of Charlemagne. Farnsworth sees an imposed heroic connection with patriarchal incest. Farnsworth writes almost at the same time that Freud consigns a daughter's claims of parental rape to hysteria. In Farnsworth's writings, incest seems to be a heroic prize for converting to patriarchy. [42]

Farnsworth's critic concedes that the Roman patriarchal system "diminished in strength as the Empire weakened." [43] Robinson seems sympathetic to Farnsworth's suggestion that there is a "hegemonic shift in the poems toward the end of the time period" (XIth to XIIIth centuries). [44] Both writers agree that "European matriarchal values survived in literature like *Beowulf* and the *Chansons de Geste*." [45] Farnsworth also refers in passing to *Beowulf*, and mentions the fact that King Hygelac was the hero's maternal uncle. [46] We may grant validity to the arguments expressed by both writers if we acknowledge that: a) the Oral Epic goes back to an earlier, more primitive age, and that b) comments on the matriarchal tradition should be seriously considered since they reveal important anthropological views. The central argument can be applied to our *Chanson* thusly: a) desire for Gile means matrilineal descent prevails; whereas b) pushing incest toward the paternity issue means the patriarchal perspective prevails. Assuming as supposition Charles' closeness to Gile is enough for our argument. We acknowledge that Robinson could be right in asserting that Farnsworth goes too far by suggesting incest, since such a notion goes beyond the native oral tradition prevalent in the Middle Ages, and makes ancient Teutonic custom subservient, through Freudian theory, to a "re-invigorated Roman empire,"

---

[42] Robinson 38.
[43] Robinson 31.
[44] Robinson 38.
[45] Robinson 22.
[46] Farnsworth 219.

pervasive before the XIth an̄ after the XIIIth century, what Robinson calls *Rome 2*.[47]

In history and in the poem Charles' virility is eminen̄. In our *Chanson*, after the execution of Ganelon, Charlemagne is eager to bring baptism to Marsile's wife, Bramimonde; she becomes the new Julianna [3978-3987]. As final symbolism of the epic poem, the sacramental rebirth of Bramimonde through baptism serves as consolation for the death of Aude, Roland's betrothed. The Christian conversion of Marsile's widow leaves us with the feeling of successful ritual. In dramatic terms, the power of the Saracens against Charlemagne has been barely enough to retain Saragossa; they lose in the end to a superior force, whose advantage is not just military but cultural as well, and which persistently wages war, albeit handicapped by an assault against royal authority at home. The Christian doctrine of regeneration through death and resurrection never aggrandizes Charles' forces much beyond the goal of future self-righteous struggle, as seen in the ending, when St. Gabriel assigns a new mission to the tired king [3993-3998]. Resolution of the authority crisis is geared toward prevention of future loss in the realm. The Christian spirit identifies with Christ's sacrifice by not losing hope in the long run, for His death brought redemption. In the literary context, we are left with the feeling that the war is far from over; through the open ending, our sympathy for Charlemagne is extended beyond the text. Moreover, although Charles may seem an invader, he is claiming back lands which were not originally "heathen." The present must be referred to the past and the future because the welfare of the realm is at stake.

Ganelon as Scapegoat

Extra-textual motivation for Ganelon's treason, central to the plot in *Roland*, may ensue from his precarious position in a patriarchy as estranged second husband to our hero's mother and the Emperor's widowed sister. His

---

[47] Robinson 24.

treachery leads to Roland's martyrdom. As the Biblical Judas, the villain becomes a crucial figure in a Christian plot. Once we rise in comprehension beyond the overt confrontation of Christians against Saracens in *The Song of Roland* we should focus on the internal antagonism prevalent within the family unit which heads the French camp. Unlike Sigmund Freud's *Oedipus Complex*, the real father does not serve as model here, nevertheless, every relation in this patriarchy has paternal and filial coloring. [48]

In the narrative Ganelon absorbs all the blame. At his trial, in his defense, the famed traitor attempts to assign to Roland complete responsibility for the risky plight he found himself in at the enemy camp:

> 'His nephew Roland, hating me in his heart,
> Had me condemned to torment and sure death:
> I was to bring Charles' message to Marsile –
> I had the wit and wisdom to survive.
> I faced Count Roland and challenged him aloud,
> And Oliver, and all the other peers.
> Charlemagne heard me, so did these noble lords:
> I am avenged, but not by treachery.'
> ....
> *'Rollant sis niés me coillit en haür,*
> *Si me jugat a mort e a dulur.*
> *Message fui al rei Marsilïun;*
> *Par mun saveir vinc jo a guarisun.*
> *Jo desfiai Rollant le poigneor,*
> *E Oliver e tuiz lur cumpaignun;*
> *Carles l'oïd e si noble baron.*
> *Venget m'en sui, mais n'i ad traïsun.'* [3771-3778]

Paramount in the Count's remarks is Roland's relationship to the king. Since Charles' nephew had nominated him as emissary on the Embassy, Ganelon is resentful enough to assign ill will to his stepson. Yet it was not all Roland's doing. The verb *jugat* [3772] denotes in meaning the nomination of someone for specific

---

[48] Girard mentions how feelings of "imitation, admiration, and veneration" may change, through the mimetic nature of desire, into the negative sentiments of "despair, guilt, and resentment." Violence 182, 188.

duties. Both protagonists are rivals. Even if Ganelon's resentment should account for personal vengeance, it could not justify rationally, nevertheless, military high treason; blind to consequences. Ganelon never mentions the twenty thousand dead at Roncevaux beside Roland. Ganelon assumes that, since there was a boast in the presence of the king [326], vengeance through treason should explain, and even justify, the great dishonor of suffering a challenge to his own paternal authority. The dreadful plot unravels itself within the family circle.

Blame against Ganelon pours on irremissibly. At the traitor's trial, Pinabel's intercession already indirectly suggests the execution to follow. The champion addresses the ill-fated Count:

> 'If any Frenchman decides that you should hang,
> The Emperor Charles must have that judgment tried:
> My sword shall prove these accusations lies.'

> '*N'i ad Frances ki vos juget a pendre,*
> *U l'empere le noz dous cors en assemble,*
> *Al brant d'kacer que jo ne l'en desmente.*' [3789-3791]

The term *juget*, as Pinabel uses it, can only mean to convict or condemn. There *should not* arise anyone brave enough to condemn Ganelon to be hung; yet, although the challenge is in the form of negative understatement, the unabashed suggestion is boastfully put forward that death would be forthcoming to whoever steps up in support of a stance against Ganelon. Pinabel, as champion, strikes great impact at court. The barons fear him: "They speak more softly because of Pinabel" [3797]. [49] Ganelon's ally fearlessly proclaims that: a) since Roland's death is absolute, b) no remedy can bring him back, and c) hence Ganelon should be acquitted [3803-3804]. Such faulty logic fails to take into account that the tragedy at Roncevaux was horrendous, hence no legal nor rational excuse as such is available for having caused it. The assertion constitutes an insult to the memory of the martyred Roland. Ganelon and his party have made a poor use of precedent

---

[49] *Pur Pinabel se cuntienent plus quei.*[3797]

in the course of the story. Even his earlier suggestion that the French send the Embassy to Marsile had not followed the proper use of precedent [222-227]; Roland is the one who recalls the ill-fated former ambassadors Basan and Basile [207-209]. Now the judges [3799-3804] and the barons [3807-3810] are the ones who insist at trial that Ganelon be acquitted. For them the loss of Ganelon could be insurmountable [3811-3813]. They suggest that his execution could equal Roland's sacrifice. Evidently, the challenge to Charlemagne's authority at court has reached its highest point.

Since the drive for *glory* pervades in the practice of heroism, the obvious need for transparent loyalty acquires ethical value. To forsake the need for basic loyalty destroys the social fiber; for this reason courage remains subservient to social norm. But, once bound by his position in the patriarchy, Ganelon's resentment could not be stopped, nor its effects avoided. In the epic plot, which includes motivation, courage attains fearlessness. After all, Ganelon is competing with Roland, who sets a high bar for courage. Yet, Ganelon's assertion at his trial that he lost status through Roland's nomination of him as ambassador to the Saracen camp, a risky assignment, does not justify treason [3757-3760]. To view vengeance as restitution for breach of honor means to confuse civil with criminal liability. Ganelon overstepped his role, blinded by passion generated through mimetic rivalry.

Corruption in the realm can only be resolved if there is a perfect hero. When all fails, the exception to the rule gains distinction. No one answers the challenge of Pinabel except for Thierry: "They all approve; no one will disagree/ But Geoffroy's brother, the chevalier Thierry" [3805-3806]. [50] As brother to the Duke of Anjou, Charles' standard bearer, Thierry defends Roland's memory and protects the king's sovereignty when he strikes down Pinabel. The outcome of a trial by combat represents God's decision; therefore, the Army does not hesitate

---

[50] *Nen ad celoi nel grant e otreit,/ Fors sul Tierri, le frere dam Geifreit.'* [3805-3806]

to sentence Ganelon and the thirty relatives who had provided surety to the king during the duel [3852]:

> The Frenchmen shout, 'A holy miracle!
> Justice demands that Ganelon must die,
> With all the kinsmen who came and took his side.'
> …………………………………..
> *Escrient Franc: 'Deus i ad fait vertut!*
> *Asez est dreit que Guenes seit pendut*
> *E si parent, ki plaidet unt pur lui.'* [3931-3933]

The king asks for an official verdict and the Army reiterates the sentence, for the thirty barons have become the enemies of Charlemagne. He had addressed the barons saying: "You are traitors, every one!"[3814] [51] Justice against noblemen is dispensed by the king backed by the community at large, for the former acts on behalf of the latter and vice-versa.

The roles of Charles, as a mother's brother, and Roland, as a sister's son, delineate thematic concerns that take us to the conclusion of the martial struggle. Roland as child stayed at the maternal uncle's fold. The hero, while dying at Roncevaux, recalls in his last breadth the king who brought him up since infancy, "who raised him in his house" [2380]. [52] The adolescent warrior received Durendal, his indestructible sword, directly from Charles, possibly in the knighthood ceremony, "The king himself presented it to me" [1121]. [53] The sentimental bond of nephew right becomes a way to seek stability in the realm. Farnsworth explains:

> Since motherhood is in any state of society the strongest of all
> ties, little wonder that the mother's clan assumed such importance
> in the life of the children, when Exogamy was so generally rendered
> necessary on account of the strict laws of Totemism. In primitive
> tribes of today members of the same totem are forbidden to

---

[51] *Ço dist li reis: 'Vos estes mi felun.'* [3814]
[52] *De Carlemagne, sun seignor, kit nurrit* [2380]. Farnsworth 44.
[53] '*Ma bonne espee, que li reis me dunat* ' [1121]. Farnsworth 48, 54.

> intermarry, the children are of the same clan as the mother, and
> thus the practice of tracing descent through the mother's totem
> is a natural outgrowth of marriage outside the clan. It is not
> surprising to find a hint of this practice of marrying outside the
> clan surviving in mediaeval literature. [54]

The individual may depend on being part of a family system in order to safeguard his true lineage through the mother's clan. [55] In the *Chanson*, feudal solidarity is expressed through numerous kin. Allied to avenge dishonor, Ganelon's thirty relatives bond in his defense and meet their doom. Nephew right is not the only form of allegiance, although it emerges as central to epic plot and true heroic prowess in *The Song of Roland*.

The praise for the warrior spirit takes on symbolic as well as semantic turns. Thierry is the exceptional knight, and the early phrase at the beginning of the poem, *Fors Sarraguce*, "Except for Saragossa" [6], parallels *Fors sul Tierri*, "But for Thierry" [3806], with metrical stress in the first two measures of the line. The familiar ring reminds us of the credit that goes to Charlemagne as conqueror, for instead of the defiant city on a hilltop, the exceptional soldier stands high above the rest as the king's champion. We have a clear sense now of who is in charge and the superior worth of Charlemagne is achieved by further emphasis on the concept of restriction and exception. Charles orders the executioner: "If one escapes, you're dead and put to shame" [3955]. [56] This command, framed in hypothetical syntax, recalls the *jongleur*'s statement about Charles's previous edict against infidels who refuse conversion: "If there are any who still resist King Charles, / He has them hanged, or killed by fire or sword" [3669-3670]. [57] The warrior king has support of his clan. And the epic code of ethics at the end of the *Chanson* promotes loyalty through fear of the consequences ensuing from treason. Twice the *jongleur* reminds the reader:

---

[54] Farnsworth 240-241.
[55] Farnsworth 242.
[56] '*Se uns escapet, morz ies e cunfunduz*'[3955].
[57] *S'or i ad cel qui Carle cuntredie, / Il le fait prendre o ardeir ou ocire* [3669-3670].

So one man's evil draws others in its wake.

....

Let no man's treason give confort to his pride.

Ki hume traïst sei ocit e altroi. [3959]

....

Hom ki traïst alter, nen est dreiz qu'il s'en vant. [3974]

Not simple deterrence, execution is the condition precedent to an established order.

To achieve a proper interpretation for the *Chanson* the reader should be alerted to the way in which the narrative sacrifices Ganelon at the end. Charlemagne asks for a verdict at the final trial: "Give me your judgment concerning Ganelon" [3751]. [58] Apparently, collective support for persecution is a measure intended to restore political differentiation. [59] With Ganelon as scapegoat, Charles' guilt stays in the realm of bad dreams, while the great hero attains martyrdom by refusing to blow the Oliphant in time. The fact that Ganelon is characterized early on in the narrative as traitor shows that the *jongleur* was reciting a tale everyone knew beforehand as part of an epic oral tradition; before the Embassy to the enemy camp, the villain appears in central stage, "Ganelon came by whom they were betrayed"[178]. [60] Regardless of oral legend, what sparks the actual event of Ganelon's betrayal in the *Chanson* is Roland's nomination of his stepfather for a deadly mission.[61] Here is the textual motivation for treason in *The Song of Roland.* Ganelon immediately in the text puts forth the oath: "If God should grant that I come home again, / I won't forget – and you'll face such a feud / That it will last as long as you're alive" [289-291].[62] His public

---

[58] *'De Guenelun car me jugez le dreit!'* [3751]

[59] Scapegoat 12.

[60] *Guenes i vint, ki la traïsun fist.*[178]

[61] "I name," says Roland, "Stepfather Ganelon." *Ço dist Rollant: 'Ço ert Guenes, mis parastre.'* [277]

[62] *'Se Deus ço dunet que jo de la repaire/ Jo t'en muvra un si grant contraire/ Ki durerat a trestut tun edage'*[289-291]. Terry wonders: "Just why Ganelon so hates Roland is not known to us; it may have been to the poet's contemporaries … A heroic stepson might well inspire a

persona seems tarnished beyond repair. Suffering from a breach of honor, he turns into a deadly enemy. The Count seeks "a little trick to play" [300]. [63] Although Ganelon's involvement in the plot is sly and non-heroic, Roland is overtly responsible for placing his stepfather in an estranged position.

Ganelon's relationship to Charles and Roland allows us to cast further light on the rationale for treason. Through warped logic, the hero's stepfather hopes to return safely home from the Embassy with two problems in the patriarchy resolved: a) Roland is removed as obstacle to patrimonial inheritance for Baldwin, Ganelon's son and the hero's half-brother; and b) the Count avoids feeling shame over a strained role as stepfather. Before his departure to the Saracen camp as Charles's emissary, Ganelon names Baldwin as his sole heir, should he die on the risky mission [313-315]. Charles answers with the famous line: "You have too soft a heart." '*Trop avez tendre coer.*'[317]

Roland retains his role as heir through matrilineal descent after his death. Taking up the warrior role to avenge his slain nephew, Charles faces his monster double in the encounter with Baligant. The execution of Ganelon, after the trial by combat between the victorious Thierry and Pinabel, establishes the king's dominance with the image of Roland looming overhead. Roland, Charles, and Thierry are morally victorious. Roland's exceptional heroism, Charlemagne's willing trust despite warning visions, and Ganelon's treason reveal a shared guilt. In the epic context we sense a drive toward royal supremacy and attainment of *glory*. Not only at Roncevaux [2415], but also during Pinabel's confrontation at the end, the Army weeps for Roland [3870-3871]. These are parallel sentiments of grief reflected in the ambiguous symbolism evident in the second dream of the first sequence. In *The Song of Roland*, the retreating Army represents an absence of participation in military agression at a crucial moment in the plot. The Frankish

---

particularly virulent jealousy, all the more acute in that it would have to be, in the case of Charlemagne's nephew, quite well concealed." Terry 11, 14.

[63] '*Einz i frai un poi de legerie.*'[300]

host displays a weak role. The Army must share the blame for not being at Roncevaux to aid the rearguard.

We find a strange counterpoint to Christian guilt in the Saracen destruction of their idols [2587-2588]. Bramimonde goes as far as to accuse the pagan gods of treason: "We are betrayed, abandoned by our gods" [2600]. [64] Abandoning faith in her gods is a positive sign, since the conversion of Marsile's wife represents the acting out of unconditional surrender. Besides overcoming Saracen forces, Charles emerges triumphant in mimetic rivalry against Marsile, with Bramimonde as object of desire. Repeatedly, a shared object of desire affects relations among multiple characters, both friends and foes, in the *Chanson*.

We marvel that a primitive scheme of suppressed desire should permanently influence our epic genre and extend into a cultural norm for sacrificial ritual. To impose his socially acquired patriarchal ties, heedless of military bondage, Ganelon goes to the enemy, and pays with his life for Roland's death. Kinsmen through matrilineal descent, Charles and Roland do not succumb to such lack of differentiation, mindful of their loyalty to France and their allegiance to a matrilineal patrimony.

Our conceptual analysis discloses a crisis caused by rivalry originating in hidden, jealous resentment over an absent figure of sister, wife, and mother extrinsic to the text. Charlemagne, Ganelon, and Roland are tormented by the need to protect posterity and end war; yet we perceive that in plot development they are responsible for the escalation of violence. All three heroes share guilt for the ambush at Roncevaux; and, irrespective of apparent intentions, consequential battles rage on and war is not over. In *The Song of Roland* the dreams of Charlemagne, the treason of Ganelon, and the death of Roland reveal a quadrangular family drama of epic proportion due to the tortured pathology of embittered heroes.

---

[64] *"Li nostre deu i unt fait felonie.* [2600]

Works Cited

Auerbach, Erich. *Mimesis: The Representation of Reality in Western Literature.*
Princeton U P, 2003.

Braet, Herman. "Le second rêve de Charlemagne dans la *Chanson de Roland.*"
Romanica Gardensia 12, 1969, pp. 5-19.

Bowra, C.M. *Heroic Poetry.* Macmillan & Co. Ltd, 1952.

Charlemagne, Wikipedia, *The Free Encyclopedia*, 6 March 2024,
http://en.wikipedia.org/wiki/Charlemagne#Appearance

Duggan, Joseph. "The Generation of the Episode of Baligant: Charlemagne's Dream and
The Normans at Mantzikert." *Romance Philology* 30, 1976-1977, pp. 59-82.

Farnsworth, William Oliver. *Uncle and Nephew in the Old French Chansons de Geste, A
Study in the Survival of Matriarchy.* Columbia U P, 1913.

Gamba, Bartolommeo. *Li Reali di Francia.* Tipografia di Alvisopoli, 1821.

Girard, René. *The Scapegoat.* Translated by Yvonne Freccero. The John Hopkins U P, 1986. -
-------*Violence and the Sacred.* Translated by Patrick Gregory. The John Hopkins U P, 1979.

Guessard, M.F. and Michelant, H. *Les Anciens Poétes de la France.* 1859 Kraus Reprint, Ltd.
1966.

La Chanson de Roland. Edited by Joseph Bédier. Vol 2. L'edition d'Art H. Piazza, 1968.

La Chanson de Roland. Edited by T.A. Atkinson Jenkins. D.C. Heath and Company, 1924.

La Chanson de Roland. Edited by Gérard Moignet. Bibliothèque Bordas, 1969.

Lévi-Strauss, Claude. *Structural Anthropology.* Translated by Claire Jacobson & Brooke
Grundfest Schoepf. Basic Books, Inc., 1963.

Owen, D.D.R. "Charlemagne's Dreams, Baligant and Turoldus." *Zeitschrift für Romanische
Philologie* 87, 1971, pp. 197-208.

Paris, Gaston. *Histoire Poétique de Charlemagne.* 1905. Slatkine Reprints, 1974.

Robinson, Marsha R. *Matriarchy, Patriarchy, and Imperial Security in Africa.* Lexington
Books, 2012.

Semple, Benjamin M. "Recognizing Roland: The Response of the Medieval Audience to the
Dreams of Charlemagne in the *Song of Roland.*" In *Dreams in French Literature: The
Persistent Voice.* Edited by Tom Conner. Editions Rodopi, 1995.

Steinmeyer, Karl-Josef. *Untersuchungen zur allegorischen Bedeutung der Traüme im
altfranzosischen Rolandslied.* Max Hueber Verlag, 1963.

The Nibelungenlied. Translated by Cyril Edwards. Oxford U P, 2010.

The Song of Roland. Translated by Patricia Terry. The Bobbs-Merrill Company, 1979.

Van Emden, W.G. "Another Look at Charlemagne's Dreams in the *Chanson de Roland.*"
French Studies 27, July 1974, pp. 257-270.

Whitehead, Frederick. "Charlemagne's Second Dream." *Olifant. A Publication of the Société
Rencesvals. American-Canadian Branch* 3, March 1976, pp. 189-195.

Mio Cid, Noble Warrior Lord

The plot of the Spanish epic of *Mio Cid* revolves around the personality of the protagonist. Along with Ramón Menéndez Pidal, we view Rodrigo Díaz de Vivar as a figure of history and literature. [1] Although the Spanish epic may lack the Holy War theme prevalent in *The Song of Roland*, Pidal maintains that the Cid exemplifies loyalty and patriotism, not the selfish attainment of wealth and glory. [2] Cesáreo Bandera Gómez affirms that there is a religious tone in the *Poema*, different from the conventional role of Crusader, which overcomes the notion of a Cid motivated by self-aggrandizement. [3] We shall explore the moral character of the Spanish hero.

The central issue supporting the notion of the Cid as champion of noble ideals is the fact that the hero foregoes the right, granted to him by statute, to wage war against the king who banishes him unjustly. [4] As he grows increasingly powerful by conquest, the poetic Cid sues for redress in three consecutive embassies, sending gifts to the king. [5] Showing religious zeal, the historic Cid

---

[1] Cantar 29.

[2] Gerald Brenan considers the Cid a mercenary. George Tyler Northup accounts for a weak religious motif because the Moors at the time were admired in Spain, and not feared. De Chasca 148-149, 156.

[3] Bandera Gómez declares that the notion of a materialistic Cid is outmoded and useless. Bandera Gómez 49.

[4] De Chasca quotes from the *Fuero Viejo de Castilla* and *Las Partidas*, cited by Menéndez Pidal, stating that the banished had a right to battle the king, and, moreover, that those vassals serving under an exiled lord had the duty to aid him in armed rebellion. De Chasca. 151. For a contrast between the historical figure who fed the legend, and the poetic hero, we rely on the works of Ramón Menéndez Pidal, considered a foremost authority on the particular epoch in which the Cid lived. De Chasca, 154. Although the Cid of history was banished twice, and was not pardoned the second time, the Cid of the epic is exiled once; only to be gloriously vindicated by a monarch whose character undergoes a gradual change toward benevolence, which probably never occurred in reality. The austere severity of the historical Cid gives way to the magnanimity and loyalty of the poetic Cid. De Chasca 145-146.

[5] Hamilton and Perry. 225. The Cid sends the king thirty horses [816], then one hundred [1274], and, finally, two hundred horses. [1813] Michael 135, 168, 202.

establishes a cathedral in the conquered city of Valencia, his preferred citadel.[6] Edmund De Chasca, along with Menéndez Pidal, believes that Mio Cid remains loyal to the king, his persecutor, in history and in the epic, because a personal pardon represents national unification.[7] We lean on this perspective to view the Cid as a victim in need of respect and consideration.[8] Since public image of the legendary figure is fostered mainly by the epic poem, we concentrate our primary exploration of the hero's character on the views expressed in the *Poema*. We cannot neglect, however, significant overlaps between poetry and history which cement the cultural conception of the Cid's identity as a heroic figure. We intend to explain why the general fascination persists through the ages.

The conquests of an exile, loyal to the king who banishes him, form a basis for the conceptual progression of *Mio Cid*. To penetrate the poetic symbolism implied in the quest for pardon means to recapture the original experience of the Spanish epic. To condemn the Cid as cruel warrior lord is inappropriate because the historical figure that fed the legend lived at a time when the political and economic progress helpful to national security depended on territorial conquest and proper government. Attaining victories against the Arab invaders, the exiled *infançón* merits better treatment from the monarch; but the nobility of long standing impede such progress through their slander and calumny of the heroic Cid. In the *Poema*, King Alfonso, eventually persuaded by the Cid's continuous conquests, intends to get the warrior back to the fold; consequently, he aids in the implementation of a matrimonial alliance with two local nobles. The unfortunate beating of the Cid's daughters by the undeserving grooms, shortly after the unsuccessful reconciliatory marriage, is the climax that brings legal scruple to bear on the difference between the hero and his enemies. As De Chasca states: "The Cid's refusal to rebel against the king, persistently seeking justice for the

---

[6] Hamilton and Perry 228.
[7] De Chasca 149.
[8] De Chasca 151.

Outrage against his daughters through legal proceedings, never resorting to unruly personal vengeance, is the best argument to defend his claim to glory." [9]

From the *Poema* we may extract the conceptual essence of morality; for the moral character does what he must, regardless of whether he gets a reward or not. Lack of recognition, even suffering, does not perturb performance of duty. The epic tradition in the Western World embraces this essential concept of morality as endurance in the wake of loss. In the *Iliad* Achilles loses Patroclus and returns to the fighting.[10] Enduring shipwreck and the loss of his men, Odysseus returns to Ithaka and slays the suitors of Penelope in the Homeric epic. [11] Neither the Platonic Socrates, nor the Biblical Christ, shun duty at the prospect of loss, rather nourishing by their paradigmatic life, whole systems of thought and feeling.[12] Charlemagne suffers the loss of his nephew in *The Song of Roland*, and steps over into legend through future conquest.[13] Mio Cid suffers exile and the beating of his daughters by the hypocritical grooms, yet remains loyal to the monarch, an *exemplum* of the moral warrior who fuels through sacrifice the drive onward with the Reconquista.[14]

The noblemen who behave ignominiously, distrusting the Cid, forfeit their noble heritage, which remains a prize to be obtained by the truly noble, an outcome to be decided in the course of the epic. The role of *infançón* makes the Cid liable to be targeted; his social rank places him below the nobility of long standing. The ranks of the Spanish nobility were basically three: in descending order, "*ricos omnes* (consisting of *condes* and *podestales*), *yfançones*, and the *fijos d'algo*, which included *caballeros* and even *escuderos*, and more widely, all men

---

[9] De Chasca 150. This translated paraphrase is mine.
[10] Lattimore's *Iliad* 353, 411.
[11] Lattimore's *Odyssey*, Book 22.
[12] Jaspers 95.
[13] Short 259.
[14] Bandera Gómez explains that "the Cid is not an idealized model, but rather a historical exemplum on how to become a true model in reality." Bandera Gómez 48. In this light the *Poema* has a theme less fabulous than *La Chanson de Roland*; the French hero appears more epic and mythical than the Cid. Bandera Gómez 71-73.

of good lineage." [15] Not being in the lowest rank, the Cid combines the marginality of the outsider with the marginality of the insider.

The exiled Cid must establish his superior worth as an *infançón* whose noble character far exceeds any conduct displayed by the false nobility. The king gradually realizes how different the Cid is from himself:

> "I sent the good Campeador into exile, and
> he has been doing great things on my behalf,
> while I treated him badly..." [16]

> *'Yo eché de tierra al buen Campeador,*
> *e faziendo yo a él mal e él a mí grand pro...'* [1890-1891] [17]

Throughout the narrative, in word and deed, the hero defines his attained differentiation in order to restore social order.

By remaining loyal to the king who banishes him unjustly, the Cid reveals in his character qualities of the *scapegoat*. By giving in to vengeance the Cid would lose his differentiation as victim, descending in heroic status to become like his envious enemies at court through mimetic reciprocity. Bandera Gómez exclaims that the Cid "fought like a regular medieval warrior, but never out of vengeful rivalry." [18] For the sake of proper literary interpretation, we must again note that the most prominent overlap between the historic and the poetic Cid is the reluctance of the leader to turn against the king who exiled him. Ramón Menéndez Pidal explains that the Cid renounces the right, granted to him as petty noble, to battle the lord who slighted him.[19] The historian insists that the old poem contains a Cidian line expressing historical truth: "I should not like to fight against

---

[15] Hamilton and Perry 231.
[16] Hamilton and Perry, 1984, p. 121. Translations are from this prose edition, unless specified otherwise.
[17] Michael 206. Direct quotes from the original text are from the Ian Michael edition.
[18] Bandera Gómez and Erickson 199.
[19] En torno 227.

132

my lord, King Alfonso" [538]. [20] Edmund De Chasca believes that this line expresses the most important of all moral victories. [21] The Cid stubbornly refuses to give in to mimetic reciprocity.

Once excluded from society, the differences between the victim and the persecutors are defined by what René Girard calls the "scapegoat mechanism." [22] Such perspective provides a suitable treatment for interpreting the Spanish epic of *Mio Cid*. In *Leviticus* 16:8 Aaron casts lots upon two goats; one is sacrificed to the Lord as a sin offering; the other is presented alive and released into the wilderness to serve as atonement for the sins of the people. *Leviticus* 16:10-22. [23] Girard strongly criticizes the use of the term *scapegoat* in the ritual sense only. [24] He advocates instead the concept of the *scapegoat* as "a structural principle that is absent from the text it structures." [25] Recognizing that unjust accusation leads to persecution is the *scapegoat* theme. "Persecute" is our word for fomenting unjust accusation. In the context of *Mio Cid* we should ponder over the unfair accusation, which results in banishment of the hero, as a means to interpret the text of the epic.

Identifying vulnerability of the lowly, or the physically diseased, is the first step in selecting the victim and initiating Girard's *scapegoat mechanism*. [26] The victim includes in his innocence the collective polarization in opposition to him. [27] Along with Bandera Gómez, we may consider that the Cid displays the role of an innocent victim almost incompatible with the courageous dimension expected of an epic hero. [28] Contagion between the victim and the persecutors is the second stage of the *scapegoat mechanism*. In this light St. Peter's denial is not only

---

[20] *'con Alfonso mio señor non querría lidiar'* [538]. La España 37. All quotes from the prose works of Menéndez Pidal are my translations.
[21] De Chasca 124.
[22] Scapegoat 174.
[23] Bullinger 155.
[24] Scapegoat 120.
[25] Scapegoat 121.
[26] Scapegoat 18.
[27] Scapegoat 39.
[28] Bandera Gomez 182.

excused, but expected. [29] In our discussion we see how the king becomes more noble in the course of the epic, and his courtiers less so; while the Cid acquires a true noble dimension. The final stage is the violence which may be repeated to become an integral cultural phenomenon. This last stage of the *scapegoat mechanism*, exemplified by the Crucifixion, brings redemption. [30] Viewing history, the author of *The Scapegoat* insists that the sentiment is not necessarily Christian, although it is proper to Jesus Christ. The critic considers Jesus Christ incomparable, since He does not succumb to the perspective of the persecutor; neither in a positive way, by agreeing with the executioner, nor in a negative way, by taking a position of strict vengeance.[31] Bandera Gómez explains that Christ, "instead of calling a legion of angels to destroy the culture of Satan, gives himself over into the hands of the people; he becomes the victim." [32] The tormented Christ does not give in to mimetic reciprocity.

The paradox of weakness and strength is evidenced at times by a change of status in history. For instance, early in our era, when the Christian movement was weak, Christians suffered persecution; but as time went on, and Christianity became strong, under Constantine, Christians eventually became persecutors themselves.[33] Jesus Christ is different from Christians. Divine power should not aim at destruction or exclusion, it should not expel anyone through, or by, persecution. Satan's power is destructive; whereas Jesus Christ brings the world a stubborn peace. [34] Evidence for this last stage in the *scapegoat mechanism*, when it occurs, is difficult to recognize as necessary to resolve a sacrificial crisis, since immediacy blurs the long range revelation of the process, or, as Girard states: "transcendent qualities are replaced by the justification of social utility." [35] The cultural anthropologist explains further that: "the political reason is always

---

[29] Scapegoat 105.
[30] Scapegoat 111.
[31] Scapegoat 126.
[32] Bandera Gómez and Erickson 204.
[33] Scapegoat 204.
[34] Scapegoat 191-192.
[35] Scapegoat 113.

contested by its victims and denounced as persecution even by those who would unwittingly resort to it should they find themselves in a position similar to Caiaphas." [36] In the *New Testament* the High Priest arrives at the decision to put Jesus to death since "it is expedient for us that one man should die for the people and that the whole nation perish not" [*John* 11:50]. [37] A savior represents the community, and does not concede victimization; nor are the persecutors willing to accept their role. Violence provides resolution to this tense denial. Bigomil Straczek considers the Caiaphas principle central to the *scapegoat mechanism* because the community is subjected to the belief that there is life in death, "sacred hierophany." [38]

The Cid retains the posture of innocent victim without relinquishing his stature as fearless warrior because he wages war against foreign invaders and ignoble courtiers. [39] When the Count of Nájera remarks that in the Cid's territory there is no Moor alive, the king responds by telling the count that the hero "serves me better than you do" [1348-1349]. The magnanimous Cid liberally shares acquired booty with his men [847-849]. While waging war, the Cid mingles charity with military strategy. He frees the captured Count Berenguel of Barcelona, who had previously denied passage to his army [1040]. Revenge may not advance safe-passage in the future. Even while engaging the enemy during warfare, the noble warrior displays, in tactical maneuvers, the behavior of the prudent. The Cid deceives the defenders of Alcocer into leaving their stronghold in pursuit of some regiments that feign defeat; the strategy to follow is to draw the enemy out after apparent retreat, "In the guise of the discreet to lure them out into an ambush." [40]

---

[36] Scapegoat 113.
[37] Bullinger, 1548.
[38] Straczek 51-52.
[39] There are issues of cruelty to be answered by all sides. Menéndez Pidal tells how the historical Cid "burned prisoners alive or subjected them to being torn apart by dogs." La España 40. Apparently, the *juglar* does not use material in the *Poema* that could injure the living legend in order to frame a figure that delineates the favorable aspects of the Cid's heroism for posterity.
[40] *a guisa de menbrado por sacarlos a çelada.* [579] The translation above is mine. The condensed poetic beauty of this simple line in the original text is unforgettable.

Vis à vis the authorities of the realm, the Cid retains the posture of victim. The king is swayed by his courtiers, but he too displays initial anger at Cidian prowess. Anthony Zahareas illustrates: "The Cid of legend and history was exiled for two reasons: one, through the calumny of his enemies and two, because Alfonso became angry after the Cid attacked the Moslem king of Toledo without his consent." [41] His struggle is twofold. The nobility cherishes belief in the culpability of the Cid as victim. The king and the nobility fuse in a representation of communal authority. The combined mimesis generates a sacrificial crisis which leads to expulsion of the victim from society. [42]

Just as banishment isolates the Cid, the forsaken nuptials separate his daughters from the parental fold. They too become victims. The noblemen unwittingly absorb the projected retribution that shall purify society. They become themselves the logical target for communal anger. In the *Poema* the *Afrenta* cannot be verified historically; we must understand it as sheer poetic contrivance. Since the outrage against the daughters is totally fictitious, we must account for it in the framework of a justifiable artistic construct. In the *Poema* Diego and Fernando González are the nephews of Count Nájera; although they belonged to Alfonso's *schola regis*, they had no right to the title of *infantes*. But this is the title the *juglar* assigns to them. Moreover, Ian Michael explains: "There is no record of their betrothal or marriage to the Cid's daughters." [43] The poetic fiction probably emanates from a legend created to signal unfair nepotism and disorder in the realm. In the course of the narrative, the grotesquely dishonorable action by the *infantes* stands out as unworthy of high lineage. We may add, as contributing effect to such dramatic character contrast the entire episode of the lion; both *infantes* cowered before the escaped lion, which the fearless Cid led back to its cage [2278-2301]. The confused *infantes* even claim the infamy of the *Afrenta* is

---

[41] Zahareas 170. Rolando Pérez, citing Bernard Reilly, explains how the Cid warped Alfonso VI's "plan to instigate rivalries between the different Muslim factions of Toledo to weaken the city's nodes of power." Pérez 127.

[42] Scapegoat 189.

[43] Hamilton and Perry 228.

justified by their need to appease the rancor they harbor, fearful that anyone may harp over their cowardice during the lion episode [2548, 2556]. Consequently, the outrage of the *Afrenta* must be accounted for within the narrative in a formalistic way. The anthropological perspective is sensible: two brothers commit a crime against two sisters, a classic case of mimetic rivalry and reciprocal desire unleashing violence.

The framework of corruption at court becomes obvious enough to reveal the overt injustice perpetrated against the victim. However, impressed by the Cid's victories, the king becomes disposed toward pardoning the Cid; the monarch hopes to honor him further with an advantageous marriage. Since the promise of pardon includes the request for double betrothal, a self-conscious vassal could hardly separate one from the other [1905-1906]. Indeed, the special assembly summoned to implement the pardon, the *Vistas*, shall be held with the prospect of the marriage already looming overhead as an implicit contingency. [44]

The Cid of the *Poema* accepts the nuptials with gratitude for the pardon, yet with distrust toward the marriage:

> "I should not wish for this marriage
> but, as the King, our overlord, urges it so strongly,
> let us discuss it quietly among ourselves."

> *'d'este casamiento non avría sabor,*
> *mas pues lo conseia el que más vale que nós,*
> *fablemos en ello, en la poridad seamos nós.'* [1939-1941]

The Cid's misgivings about the marriage are based largely on the *infantes'* haughtiness and the connections they hold at the royal court: "They are very proud young men and they belong to the King's household" [1938]. [45] The rivalry between the Cid and the dissemblers at court, or *mestureros*, gives way to the

---

[44] "There were three kinds of gatherings, in ascending order of importance: *juntas* ('meetings'), *vistas* ('assemblies'), *cortes* (convocations of the royal court or *curia regia*)." Hamilton and Perry 231.
[45] *Ellos son mucho urgullosos e en part en la cort.* [1938]

higher consideration of obedience to the *rex imago*. Menéndez Pidal explains: "Those so called dissemblers, or minglers (that is to say, sowers of discord) constituted a true public calamity which deeply destroyed social life." [46] What is rendered as "overlord" in Hamilton and Perry's translation, actually, in the original text, means "the one who is worth more than us," *el que más vale que nós* [1940].[47] The humility of the Cid emerges as a political stance. The inferior worth of the grooms is evident; in a deplorable financial state, the *infantes* make ready for the *Vistas* by pawning goods, "they paid cash for some things and obtained others on credit" [1976].[48] The decadent nobility does not possess much more than the common populace. The conquering Cid, in turn, displays diffidence to the king. The hero, upon encountering the monarch, prostrates himself, pledging devotion. The king tells the Cid to rise on condition that, if the hero does not behave more normally the monarch will withdraw his love [2027-2029]. The humble hero insists that he shall not rise until he has *amor* from his *natural señor* loud enough for all to hear [2031-2032b]. The warrior lord's province of Valencia seems to become officially annexed when the king ratifies the governorship by granting the Cid a place in his kingdom, along with the pardon:

> The King replied: "I shall do so with all my heart.
> Here and now I pardon you and restore you to my favour
> and welcome your return to my kingdom."
>
> *Dixo el rrey: "Esto feré d'alma e de coraçón;*
> *aquí vos perdono e dovos mi amor*
> *[e] en todo mio rreino parte desde oy.'* [2033-2035]

Acknowledging the Cidian territory of Valencia as part of his kingdom the king dispossesses himself indirectly. In this obvious case of contagion, the donor

---

[46] De Chasca 69.

[47] Hamilton and Perry 124-125.

[48] *lo uno adebdan e lo otro pagavan* [1976]. The Cid confronted impecuniosity in his exile. Upon his arrival at Burgos the Cid "confesses that he doesn't have any money with which to feed his men." Pérez 130.

makes the receiver another self. The treaty joins both characters through mutual demands since allegiance and duty are involved.

This pardon scene would represent a climax in the story, except for the betrayal and subsequent retribution in the *cortes* and the duels. Mindful of his role as Christian warrior, loyal to the king as institution, the Cid's acceptance of the long-sought pardon is directed to God, to Alfonso, and to his army:

> "I receive your pardon with gratitude, my lord Alfonso.
> For it I thank God, then you
> and these my vassals who stand here with me."
>
> *'¡Merçed! Yo lo rreçibo, Alfonso mio señor;*
> *gradéscolo a Dios del cielo e después a vós*
> *e a estas mesnadas que están aderredor.'* [2036-2038]

The moral strength of the Cid justifies his appeal as military hero for the ages. Menéndez Pidal holds that "The noble ethics of the exile from Vivar was, then, one of the main reasons why he became the object of legendary song." [49]

The morning after granting the long awaited pardon the king requests that the marriages be effectuated:

> "I ask the hands of your daughters, Doña Elvira and Doña Sol,
> In marriage for the Infantes of Carrión."
>
> *'Vuestras fijas vos pido, don Elvira e doña Sol,*
> *que las dedes por mugieres a los ifantes de Carrión.* [2075-2076]

We see that, before leaving the king's presence, the heroic Cid insists that it is the monarch himself, and not he, who marries the girls, "it is you, not I, who are giving my daughters in marriage" [2110].[50] Moreover, back in Valencia, the Cid will announce the marriage to his family with a disclaimer of liability as caveat:

---

[49] De Chasca 57.
[50] *'Vos casades mis fijas ca non gel as do yo.'* [2110]

139

"it is he who is giving you in marriage and not I" [2204].[51] The hero's detachment from the betrothal maintains his established superior worth, while his acceptance of the marriage shows subservience to the king's will. Although the monarch views the pardon as intimately entwined to the betrothal, the worried father wishes to distinguish the two. Lack of such differentiation is foreboding.

Acceptance of the marriage, combined with recalcitrance over accepting agency for the same, outlines the stern difference between the Cid and the *infantes*. The line describing the separation of the daughters from the parental fold duplicates the previous torn nail metaphor, which was used to describe the earlier separation [375] of the Cid from his wife, Jimena, and the family: "They parted with such pain as when a finger-nail is torn from the flesh" [2642].[52] The dramatic echo draws a conceptual parallel between the *mestureros*, responsible for the Cid's exile, and the ungrateful *infantes*; cruelty and greed outline inferior worth for both factions, causing suffering which the Cid and his family must bear; for the bonds of matrimony are to be severed by a nefarious breach. Yet, unlike the subtle intrigue at court that triggers the unjust exile, the personal outrage of the cowardly *Afrenta* is openly criminal. [53]

As Doña Sol, one of the Cid's daughters, warns, the wretched deed will stand as testimony for the worthlessness of the *infantes*:

> "Do not ill-treat us like this,
> for if we are beaten you will be disgraced and
> men will charge you with this crime in assemblies and courts of justice."

> *'Atan malos ensienplos non fagades sobre nós.*
> *si nós fuéremos maiadas, abiltaredes a vós,*
> *rretraer vos lo an en vistas o en cortes.'* [2731-2733]

---

[51] *'bien me lo creades que él vos casa, ca non yo.'* [2204]

[52] *Cuemo la uña de la carne ellos partidos son.* [2642]

[53] Girard insists that, beyond a certain threshold, hate exists without a specific reason. Scapegoat 38.

Even though Doña Sol concludes her plea for mercy with a reminder of the impending suit, the *infantes* remain heedless. The king could not possibly condone the action since the *infantes* behave as if the Cid were still the exiled *infançón*. They do not heed the acquired differentiation, merit attained through noble deeds. The corrupt nobles' beating of the Cid's daughters is a monstrously unfair event in plot development; the *infantes* were granted the role of grooms by an overzealous monarch. [54] Alfonso had implemented the nuptials in the hope of attaining unity of classes. But, instead of aiding the ascension of the Cid's stature, the nobles are shown to be worth less than the Cid, projecting further contagion in the social order.

The epic of *Mio Cid* transitions from the hero's persistent wooing of the king for the long sought pardon, through the three embassies, on to the demand for retribution for the *Afrenta*. Hence the monarch must grant redress for the egregious crime perpetrated by the *infantes*. The plot progresses into the court scene and on to the final duels which establish that the pardon and the ill-fated matrimonial alliance must converge in the medieval relief brought about through trial by combat. [55]

The failed recognition of noble value unveils the *infantes* as wretched victims themselves. [56] The cruelty of their crime forbids the reader's attachment to them; we identify them as a destructive faction, undeserving of the pity saved for the Cid and his daughters. By belonging to a lower class the *infançón* warrior and his family outline a marginality typical to the outsider; the banishment and the *Affront at Corpes* reinforce the separation. The nobles remain insiders who reflect in their spoiled character defects of the ruling cast.

---

[54] Girard explains how marriages are not clearly perceived as exchanges when society is in a state of social crisis. Institutional collapse obliterates "hierarchical and functional differences, so that everything has the same monotonous and monstrous aspect." Scapegoat 13.

[55] "Trial by Combat was, in theory at any rate, a means of causing God to pronounce a public judgement on which of the two combatants is in the right." Hart 170.

[56] Perpetrators of persecution, where the true nature of difference is not recognized, "possess the marks that suggest a victim." Scapegoat 24.

There is an element of choice which prevails prominently. The *infantes* choose to perpetrate the *Outrage* against the Cid's daughters, just as the *malos mestureros* choose to alienate themselves from the conquering warrior. The Cid does not willfully become their persecutor the way they choose him as victim. They wish to banish him from the realm. In the course of the epic the false nobility is unveiled as persecutors guilty of polluting the community. The Cid declares himself victor through the toil of his labors as conqueror and by overcoming the personal injury suffered unjustly. He deserves relief; they earn the defeat of the vanquished. Thus, differentiation is restored through their punishment, and society is saved, ending a cultural crisis. Through a sacrificial stance, the Cid attains mythological status. The *Afrenta* is perpetrated against the Cid's daughters, not due to any specific act they committed, but because "they belong to a class that is particularly vulnerable to persecution." [57] They are guiltless characters in the epic. Their innocence makes them liable, amplifying the indecent cowardice of the *infantes*. The ruffians are less wealthy, and less noble than their victims. The differences between the noble *infançón* and the false nobility is unveiled in terms of the envious hatred they display.

The *infantes* feel justified in the assault on the Cid's daughters due to their lineage. As Count Garçi Ordóñez states at the trial:

> "The lords of Carrión are of such noble lineage that
> they should not consider his daughters fit to be their concubines."

> *'Los de Carrión son de natura tal*
> *non ge las devién querer sus fijas por varraganas.'* [3275-3276]

Such condescension is appalling. The confused Don Fernando, likewise, reiterates his uncle's view:

---

[57] Scapegoat 17-18.

"We are of the family of the counts of Carrión
and have the right to marry the daughters of kings and emperors,
and the daughters of petty nobles are not our equals."

'De natura somos de condes de Carrión,
deviemos casar con fijas de rreges o de enperadores
Ca non perteneçién fijas de infançones.' [3296-3298]

The issue of greater worth by the lazy nobles is a futile groping for supremacy which their cowardly acts belie. But the trial is destined to establish justice. Per Bermúdez, the Cid's nephew, counters the false argument by an accusation of lesser worth:

"By deserting them you incurred infamy.
They are women and you are men,
But they are your superiors in every way."

'por quanto las dexastes menos valedes vós;
Ellas son mugieres e vós sodes varones,
En todas guisas más valen que vós.' [3346-3348]

Despite a difference in class, the nobility is not at par with the Cid's family in proper behavior. Even gender distinction is reversed by the traitors' cowardice. Differentiation must be established, neither at the level of lineage, nor gender, but, rather, at the level of merit, as dictated, more specifically, by proper conduct. The *infantes* lack true worth. Their argument of superior lineage is an invalid explanation; there is no justification to account for the *Outrage*. Persecutors may experience hate without cause. [58] The *Poema* reveals the nobility's decadence.

We should dwell on a crucial parallel between history and the *Poema*. Menéndez Pidal enumerates historical circumstances that support the herculean prowess of the Cid. The historian explains that Garçi Ordóñez, a favorite of the king, "practiced a lesser form of *Reconquista*, the economic weakening of Moorish lords." The inglorious Count of Nájera, uncle of the *infantes*, committed

---

[58] Girard insists on the issue of "hate without cause." Scapegoat 103, 113, 124, 146, 204.

the blunder of harassing Motámid in Seville while the Moor enjoyed sanction from the Cid as regional collector of tributes. [59] The Cid made proclamation of alliance to the Moorish king in an ultimatum the count did not heed, advancing on through to Cabra, a castle on the nearby frontier. The Cid, who had been idle for a while, launched an attack with the small host that served him as escort and overcame enemies in vast number. Although the Cid incarcerated the count merely for three days to prove the battle was won, the count's pride was injured. The Arab historians recorded it as extraordinary, an event of everlasting importance. [60] The critic proceeds: "Although the humiliation of Garçi Ordóñez pleased the people, it was very unpleasant to King Alfonso, who cherished a predilection for the Count of Nájera." [61] The difference in political tactics apparently did not affect royal favoritism; yet the unusual victory, in fact "awakened envy in many." [62] This character trait seems to have been shared by the king. Menéndez Pidal says elsewhere that Alfonso VI was considered invidious, envious, or, in the etymological sense, an adverse viewer who views with bad eyes another's merit and tries to impede it; and this invidious king gives audience and support to the crowd of dissemblers or informers, the *catellani invidentes* (envious Castilians), as they are called in the *Historia Roderici*, and he ends up by expulsing from Castile the hero, whose pre-eminence bothered him.[63]

---

[59] Menéndez Pidal explains that taxes were "a contribution which was rendered as a form of homage, that is, annual payment of tribute by the Moorish prince in exchange for protection and aid from the Christian lord." La España 84. The historian adds: "But the control through tribute was very instable. As soon as the power wielded by the Christian dominator waivered, the Moorish vassal ceased to pay, or took his tribute to another lord more powerful who threatened or entreated him; and battles raged over the *parias*: since the middle of the XIth century the system of *Reconquista* began to be superimposed to the imposition of tribute." La España 85. The Middle Ages provided the warrior spirit with an incontrovertible arena.
[60] La España 287.
[61] La España 288.
[62] La España 289.
[63] Castilla 163. The envy at court is the key point establishing differentiation between the Cid and his persecutors. Girard justifies Satan's banishment to Hell by viewing the Gospels as a way to end "humanity's imprisonment in a system of mythological representation based on the false transcendence of a victim who is made sacred because of the unanimous verdict of guilt." He asserts that this *transcendence* is called Satan, and "of all Satan's faults, envy and jealousy

144

Menéndez Pidal's exaggerated supposition could account for the exile; however, the Alfonso of the *Poema* was rather swayed by ill advice. [64] We gather that the friction was of long standing. The Cid of the *Poema* reminds Garçi Ordóñez not to reprimand him for a disheveled beard since no one has shown the hero such disrespect. The count's experience is otherwise. Apparently, the poetic Cid vigorously pulled at the count's beard. The Cid exclaims:

> "No woman's son has ever plucked it and no one, Moor or Christian,
> ever tore it –as happened to yours, Count, in the Castle of Cabra.
> When I took Cabra and plucked your beard."
>
> *'nimbla messó fijo de moro nin de christiana,*
> *commo yo a vós, conde, en el castiello de Cabra;*
> *quando prís a Cabra e a vós por la barba."* [3286-3288]

The Cid's reference to the incident at the Cabra castle, in which his will to favor the Moorish king predominated over Garçi Ordóñez's greed for more tribute, shows the hero's superior worth [3287-3290]. Evidently, in history and in the *Poema*, a confrontation establishing supremacy occurred at the Cabra castle.

In the *Poema* the main drive toward superior worth is outlined by a comparison between the hero and the monarch. Ian Michael considers the *pensó e comidó* line, meaning 'thought and deliberated,' "a formula used for the receipt of bad news." [65] The textual critic refers to several uses: 1) first employed when the king hears the *infantes'* request for the betrothal [1889]; 2) and used as well when the Cid receives from his ambassador, Minaya, news of the long awaited pardon, mingled with the marriage proposal [1932]; 3) the formula occurs also as the Cid hears about the infamous *Afrenta* [2828].[66] In addition to these instances, 4) the formula is repeated when the king receives the Cid's request for justice for

---

are the most in evidence; Satan could be said to incarnate mimetic desire were that desire not, by definition, disincarnate." Scapegoat 166.

[64] En torno 141-144, 158.
[65] Hamilton and Perry 231
[66] Hamilton and Perry 231

the *Outrage* [2953]. Both the Cid and the monarch share an identical display of sullenness during the intense deliberation proper to those in a true position of authority.

The final scenes of the epic overcome tragedy and insult through the dispensation of justice. Yet, after the trial, the Cid departs. The final retribution is left to the Champion's retainers. Mio Cid is not present during the duels that restore justice. [67] His absence displays rightful indignation, while conveying the fact that the hero does not give in readily to mimetic reciprocity. The differentiation between the Cid and his enemies is absolute. [68] He departs as gesture of disdain; he has agents to leave in charge; the king himself gives the measure of security, "as a lord does for a good vassal" [3478].[69] Although we miss here the adjective "good" with "lord," the earlier plaint of the people of Burgos seems answered, "What a good vassal. If only he had a good lord" [20]. [70] The adverbial conjunction "if" in the famous 20th line, outlines surprise, rather than expressing a true hypothesis; the clause expresses regret that there should be need of a better lord. By not warring against the king, the Cid is already a moral subject, "good vassal." [71] The hypothetical use of "if," *si*, as an adversative adverbial

---

[67] To regain his honor was a strong motivation for the Cid's heroism. His honor was compromised by his banishment and by the abuse of his daughters. In both cases he regains his honor through battle. Stephanie Matos-Ayala suggests that in these two instances he may combine the youthful vigor of Roland, and the authority of the elder Charlemagne. Matos-Ayala 42-45. Therefore, we may conclude that it is sensible for the elder Cid to delegate chastisement at the duels to his men.

[68] The debasement of the nobility is most obvious during the trial by combat, when Don Diego shouts out, fearing Martin Antolinez's sword: "Diego González held a sword in his hand but did not use it, / and then he (the *infante*) shouted out: / 'Help me, great God! Protect me from this sword!'" *Diago Gonçález espada tiene en mano, mas no la ensayava, / essora **el infante** tan grandes vozes dava: / '¡Valme, Dios, glorioso señor, e cúriam d'este espada!'* [3662-3665] Michael comments: "mention of Diego's rank makes his cowardice appear more shameful." Hamilton and Perry 240.

[69] *'yo vos lo sobrelievo commo a buen vassallo faze señor.'* [3478]

[70] *'¡Dios, qué buen vassallo, si oviesse buen señor!* [20]

[71] Miguel Garci-Gómez adumbrates the meaning of the line with delicacy: "Those curious and compassionate burgaleses, upon seeing the good vassal accept his misery patient and composed, were wishing him that he should find a good lord, who would know how to convince the monarch of the vassal's kindness, neutralizing with clemency the malignancy of the evil enemies." Garci-Gómez 77. Bandera Gómez links the line to an overview of the Epic: "In the

conjunction in the dependent clause of a conditional statement, usually indicates a necessary contingency precedent a specific result in a rhetorical framework. In his address to the lords of *Carrión* at court, before departing, the hero addresses the unrepentant *infantes* directly: "If you do not give satisfaction for this crime, let the court pass judgement" [3269].[72] In the absence of any justification for the *Outrage*, the final sentence is sure.

Irremissibly, the plot advances on to final retribution. William Entwistle explains that during the Middle Ages it was believed that in a trial by combat "victory was on the side of truth." [73] For this reason, the king closes the argument succinctly:

> "If you are successful in the combat you will gain great honour,
> and if you are defeated do not blame us."
>
> *'Si del campo bien salides, grand ondra avredes vós,*
> *e si fuére[des] vençidos, non rrebtedes a nós.'* [3565-3566]

The outcome of the duels must be accepted as final. Alfonso is adjudicating as a just monarch should, again answering the plaint of the people of Burgos.

Mio Cid combines desert for his attained glory along with acceptance for the vicissitudes of life. Three times in the course of the Epic the Cid thanks God for his misfortunes. The first instance is early on, at the start of the *Poema*:

> 'I give Thee thanks, O God, our Father in Heaven.
> My wicked enemies have contrived this plot against me.'
>
> *'¡Grado a ti, Señor, Padre que estás en alto!*
> *Esto me an buelto mios enemigos malos.'* [8-9]

The hero is grateful to the Lord for the fate dealt to him by the evil dissemblers at court. As he embarks on his exile, he feels the duty to bear his cross. The same

---

Poema are represented those two Alfonsos, the King as institution, and the King as individual."
Bandera Gómez 39.
[72] *'Si non rrecudedes, véalo esta cort.'* [3269]
[73] Entwistle 13.

147

gratitude is expressed directly to Christ upon hearing about the foreboding marriage proposal:

> 'I give thanks to our Lord Jesus Christ for this favour.

> *'Esto gradesco a Christus el mio señor.'* [1933]

The Cid equates his destiny to the divine will. His life is at the mercy of Providence. Even after learning about the opprobrious *Afrenta*, the noble warrior lord similarly thanks Heaven, addressing Christ again directly:

> 'Thanks be to our Lord Jesus Christ
> for this honour the Infantes of Carrión have done me.

> *'Grado a Christus, que del mundo es señor,*
> *quando tal ondra me an dada los ifantes de Carrión.'* [2830-2831]

The repeated address to Christ convinces the reader that the Cid considers himself a chosen vessel of God. Origin for expression of the sentiment is found in Psalms 119:71: "It is good for me that I have been afflicted; that I might learn Thy statutes." [74] The Cid's resignation in the *Poema* reveals blind faith. These invocations are characteristic of other formulaic expressions used throughout: "When his prayer was ended the Cid remounted" [54]; "When Mass has been said we must prepare to ride away [320]." [75] The various epic formulas express an obstinate submission under divine rule.

Leo Spitzer claims "historical deconstruction of a historic theme under influence of the legend;" the German critic admits that "the Cid brings to unique fruition his own being, the ideas of his epoch arise simply as emanations from his character." [76] The doctrine extracted is the spirit of the age: the charismatic leader was humble to the king, paternal to his men. As we have seen, the *juglar* remarks,

---

[74] Bullinger 840.
[75] Hamilton and Perry 25, 39.
[76] Spitzer 116.

in counterpoint measure to the king [1889, 2953], how the Cid deliberates with care before passing judgment [1932, 2828]. His legal concerns reveal that his word becomes law. Karl Jaspers reminds us that Socrates and Christ are paradigmatic individuals who make an impact on civilization by example, without writing themselves.[77] Similarly, the Cid drank the draught of envy, yet remained a pillar for posterity no less.

In the context of *Mio Cid* we view the hero's slow progression along the plot of the epic as a path toward manifestation in his character of proper attributes, without mimetic interference from envious or evil factions at court. The Spanish leader refuses to give in blindly to vengeance. The *juglar* closes the *Poema* by drawing a parallel between the Cid and Christ; besides being grateful to God for his misfortunes, Mio Cid dies the day of Pentecost [3726].[78] Pentecost is the day after Christ's Ascension, when the Holy Spirit descends upon the apostles gathered in mourning. In the last chapter of *The Scapegoat* Girard states that the Paraclete is "the destroyer of every representation of persecution. He is truly the spirit of truth that dissipates the fog of mythology."[79] Silvio Simonetti explains that Girard considers myth malignant because it reverses the roles of the mimetic victim and the persecutor; the Gospels represent the truth by impeding overlap between victim and tormentor.[80] Moral differentiation is an absolute categorical distinction. A comparison drawn between the Cid and Christ becomes plausible if we concede that the Cid is also reputed to be a historical as well as a legendary

---

[77] Jaspers 94.

[78] Pentecost is celebrated the seventh Sunday after Easter Sunday, which usually falls in late May or early June. The day is considered the start of the church's mission to the world. It is doubtful that the historical Cid died the Day of Pentecost. Ian Michael mentions that, according to the *Historia Roderici*, the Cid died in July 1099; and Pentecost fell on May 29 that year; consequently, lines 3726-3727 are considered additions by the XIVth century scribe of the original manuscript from 1201-1207. Hamilton and Perry 14, 309. We may ascribe the Pentecost-death to poetical fiction.

[79] Scapegoat 207.

[80] Simonetti blog.

figure. For Bandera Gómez the Cid is a "warrior Christ" who transcends historic reality.[81]

Symbolically, just as the Holy Spirit descends upon the apostles after Christ's Ascension to Heaven, the spirit of the *Campeador* remains a pervasive presence among the living. Perhaps the conquering Cid provided inherent ideals of persistent toil to the later *conquistadores* who sought to export the mission of Church and State to a hostile New World. They were caught in the horrendously contradictory role of invaders, bringing supposed salvation to the innocent native population, which they subdued by brute force. And yet, their violence in the New World, and subsequent conquests, triggered the Columbian Exchange, which started the commerce between Europe and the Americas that has lasted for centuries. We look toward René Girard as a writer who dwells on "the violent origins of all human culture." [82] To resolve issues of past and future historicity involved in proper interpretation of *Mio Cid* is forbidding; but actual existence of the legendary figure suggests that the unjustly banished, grieving, warrior lord may share moral fiber with the paradigmatic figures of Western civilization.

---

[81] Bandera Gómez affirms that exemplariness in the heroic Cid constantly faces the ambivalence of an uncertain future. Bandera Gómez 151, 171. Quoting from *Mythical Thought*, second volume of Ernst Cassirer's *The Philosophy of Symbolic Forms* (pgs. 105-106), our critic reminds us that in the mythical world even the past lacks a cause. Bandera Gómez 143.
[82] Violence 244.

Works Cited

Bandera Gómez, C. El "Poema de Mio Cid": Poesía, historia, mito. Editorial Gredos, 1969.

Bandera Gómez, C., & Erickson, A. "René Girard, Friendship and Battling to the End: A Conversation with Cesáreo Bandera." Contagion, Journal of Violence, Mimesis and Culture, 26, 2019, pp. 195-208.

Cassirer, E. The Philosophy of Symbolic Forms. Vol 2. Mythical Thought. Yale UP, 1955.

De Chasca, E. El arte juglaresco en el Cantar de Mio Cid. Editorial Gredos, 1967.

Entwistle, W. J. "My Cid-Legist." BSS, 7, pp. 9-15.

Garci-Gómez, M. Mio Cid: Estudios de endocrítica. Editorial Planeta, 1975.

Girard, R. The Scapegoat. Translated by Ivonne Freccero. The John Hopkins Press, 1986.

    Violence and the Sacred. Translated by Patrick Gregory. The John Hopkins Press, 1979

Hart, T. R. "Hierarchical Patterns in the Cantar de Mio Cid." RR, 53, 1962, pp. 161-173.

Jaspers, Karl. Socrates, Buddha, Confucius, Jesus. The Paradigmatic Individuals. Harcourt, Brace & World, Inc., 1962

Matos-Ayala, S. "Chivalry in Literature and Practice: A Study of the Medieval Code of Arms before and after the Arthurian Legends and its Practice in the Battlefield." 2013.Theses and Dissertations (All). 1171.http://knowledge.library.iup.edu/etd/1171

Menéndez Pidal, R. editor. Cantar de Mio Cid. Translated by Alfonso Reyes. Espasa Calpe, 1991.

-------- Castilla, la tradición y el idioma. Espasa Calpe, 1945.

-------- En torno al Poema del Cid. EDHASA, 1970.

-------- La España del Cid. Vol. I. Editorial Plutarco, 1929.

Michael, I., editor Poema de Mio Cid. Editorial Castalia, 1976.

Pérez, Rolando. "The Cid: War Machine, Self-Made Noble and Mythic Hero of Menéndez Pidal's Nationalistic Epic." Heroes and Antiheroes: A Celebration of the Cid, edited by Anthony J. Cárdenas-Rotunno. Hispanic Society of America, 2013, pp. 125-145.

Short, I., editor. La Chanson de Roland. Librairie Générale Française, 1990.

Simonetti, S. How Jesus Christ Upended the Scapegoat Myth: A Girardian Interpretation. April 18, 2019. blog.action.org/archives/108119-how-jesus-christ-upended-the-scapegoat-myth-a-girardian-interpretation.html

Spitzer, L. "Sobre el carácter histórico del Cantar de Mio Cid." NRFH, 2, 1948, pp. 105-117.

Straczek, B. "René Girard's Concept of Mimetic Desire, Scapegoat Mechanism and Biblical Demystification." Poszukiwania Naukowe. 35, 2014, pp. 47-56. 10.21852/sem.2014.4.04

The Companion Bible. Companion Bible: The Authorized Version of 1611, with Structures and Critical, Explanatory, and Suggestive Notes and with 198 Appendixes. Edited by E.W. Bullinger. Kregel Publications, 2000.

The Iliad of Homer. Translated by R. Lattimore. The U of Chicago P, 1961

The Odyssey of Homer. Translated by R. Lattimore. Harper Perennial, 1991.

The Poem of the Cid. Translated by R. Hamilton and J. Perry. Edited by Ian Michael. Penguin Books, 1984.

Zahareas, A. "The Cid's Legal Action at the Court of Toledo. RR, 56, 1965, pp. 161-172.

List of Cited Works

Adkins, A.W.H. *Moral Values and Political Behavior in Ancient Greece.* W.W. Norton, 1972.

Andrews, S.O. *Postcript on "Beowulf."* Russell & Russell, 1969.

Auerbach, Erich. *Mimesis: The Representation of Reality in Western Literature.* Princeton U P, 2003.

Bandera Gómez, C. *El "Poema de Mio Cid": Poesía, historia, mito.* Editorial Gredos, 1969.

Bandera Gómez, C., & Erickson, A. "René Girard, Friendship and Battling to the End: A Conversation with Cesáreo Bandera." *Contagion, Journal of Violence, Mimesis and Culture,* 26, 2019, pp. 195-208.

Barney, Stephen A. *Word-Hoard. An Introduction to Old English Vocabulary.* Yale UP, 1977.

Barton, Andrew. *The Knight's Progress and Virtual Realities: The Medieval Adventure from Beowulf to Ready Player One.* 2018. Texas State U, MA thesis.

Bédier, Joseph. *La Chanson de Roland.* Vol 2. *Commentaires.* L'edition d'Art H. Piazza, 1968.

Beowulf. Translated by David Wright. Penguin Books, 1968.

Bosworth, Joseph. *An Anglo-Saxon Dictionary.* Edited by T. Northcote Toller. Oxford UP, 1976.

Bowra, C.M. *From Virgil to Milton.* The MacMillan Co. Ltd., 1963.

-------- *Heroic Poetry.* Macmillan & Co. Ltd. 1952.

Braet, Herman. "Le second rêve de Charlemagne dans la *Chanson de Roland.*" *Romanica Gardensia* 12, 1969, pp. 5-19.

Cassirer, E. *The Philosophy of Symbolic Forms.* Vol 2. *Mythical Thought.* Yale UP, 1955.

Charlemagne, Wikipedia, *The Free Encyclopedia,* 6 March 2024, http://en.wikipedia.org/wiki/Charlemagne#Appearance

Chase, Colin. "Opinions on the Date of *Beowulf,* 1815-1980." *The Dating of Beowulf,* edited by Colin Chase. U of Toronto, 1981. pp. 3-8.

Clanchy, M.T. *From Memory to Written Record: England, 1066-1307.* Harvard UP. 1979.

Cunliffe, Richard J. *A Lexicon of the Homeric Dialect.* U of Oklahoma P, 1963.

De Chasca, E. *El arte juglaresco en el Cantar de Mio Cid.* Editorial Gredos, 1967.

Dictionnaire de l'Académie Française. 8 ème édition. 1987-2010. http://www.mediadico.com/dictionnaire/definition/mont-joie/1

Duggan, Joseph. "The Generation of the Episode of Baligant: Charlemagne's Dream and The Normans at Mantzikert." *Romance Philology* 30, 1976-1977, pp. 59-82.

Durkheim, E. and Mauss, M. *Primitive Classification.* Translated by Rodney Needham. U of Chicago, 1967.

Entwistle, W. J. "My Cid-Legist." *BSS,* 7, pp. 9-15.

Farnsworth, William Oliver. *Uncle and Nephew in the Old French Chansons de Geste, A Study in the Survival of Matriarchy.* Columbia U P, 1913.

Foley, John Miles. *The Theory of Oral Composition. History and Methodology.* Indiana U P. 1988.

Forney, Kristine, and Machlis, Joseph. *The Enjoyment of Music.* W.W. Norton & Company. 2008.

Fulk, R.D., et al., editors. *Klaeber's Beowulf and the Fight at Finnsburg.* 4th ed. U of Toronto P, 2008.

Fuller, Lon L. and Eisenberg, Melvin A. *Basic Contract Law.* Thomson/West, 2006.

Gamba, Bartolommeo. *Li Reali di Francia.* Tipografia di Alvisopoli, 1821.

Gans, Eric. *Chronicles of Love and Resentment 771 Origins of GA II: The End of Culture?* https://anthropoetics.ucla.edu/view/vw771/2023.

------- *Originary Thinking: Elements of Generative Anthropology*. Stanford U P, 1993.
------- *Signs of Paradox: Irony, Resentment, and Other Mimetic Structures*. Stanford P, 1997.
------- *The End of Culture*. U of California P, 1985.
------- *The Origin of Language*. A New Edition. Spuyten Duyvil, 2019
Garci-Gómez, M. *Mio Cid: Estudios de endocrítica*. Editorial Planeta, 1975.
Gierke, Otto. *Associations and Law. Edited and translated by George Heiman*. U of Toronto P, 1977.
Girard, René. *The Scapegoat*. Translated by Yvonne Freccero. The John Hopkins U P, 1986. -
------- *Violence and the Sacred*. Translated by Patrick Gregory. The John Hopkins U P, 1979.
Gordon, Cyrus H. *The Common Background of Greek and Hebrew Civilizations*. W.W. Norton, 1965.
Green, D.H. *The Carolingian Lord: Semantic Studies in Four Old High German Words: Balder, Frô, Truhtin, Hêrro*. Cambridge UP, 1965.
Greenfield, Stanley B. *"Gifstol* and *goldhoard* in *Beowulf." Old English Studies in Honor of John C. Pope*, edited by Robert B. Burlin and Edward B. Irving, Jr. U of Toronto, 1974, pp. 107-117.
-------- "Of Words and Deeds: the Coastguard's Maxim Once More." *The Wisdom of Poetry*, edited by Larry D. Benson and Siegfried Wenzel. Western Michigan U, 1982, pp. 45-51.
Guessard, M.F. and Michelant, H. *Les Anciens Poétes de la France,* 1859. Kraus Reprint, Ltd. 1966.
Guillaume d'Orange: Four Twelfth-Century Epics. Translated by Joan M. Ferrante. Columbia U P, 2001.
Hart, T. R. "Hierarchical Patterns in the *Cantar de Mio Cid." RR*, 53, 1962, pp. 161-173.
Hill, John M. *The Cultural World of 'Beowulf.'* U of Toronto P, 1965.
History of Contract Law, Wikipedia, The Free Encyclopedia, Wikipedia Foundation, 23 October 2004.
    https://en.wikipedia.org.wiki/history_of_contract_law
Homer. *The Iliad*. https://www.perseus.tufts.edu/hopper/text?doc=Perseus:text:1999.01.0133
Jaeger, Werner. *Paideia: The Ideals of Greek Culture*. Vol. 1. Translated by Gilbert Highet. Oxford U P, 1970.
Jarisinski, Stefan. *Ancient Priviliges: "Beowulf," Law, and the Making of Germanic Antiquity*. West Virginia UP, 2006.
Jaspers, Karl. *Socrates, Buddha, Confucius, Jesus. The Paradigmatic Individuals*. Harcourt, Brace & World, Inc., 1962.
Jespersen, O. *A Modern English Grammar*. Vol. 5. A. Stifsbogtrykkeri, 1949.
Jotischky, Andrew, and Hall, Caroline. *Historical Atlas of the Medieval World*. Penguin Books, 2005.
Kiernan, Kevin S. "The Eleventh-Century Origin of *Beowulf* and the *Beowulf* Manuscript." *The Dating of Beowulf*, edited by Colin Chase. U of Toronto, 1981, pp. 9-21.
Klaeber, Fr., editor. *Beowulf and the Fight at Finnsburg*. 3rd ed. D.C. Heath, 1950.
La Chanson de Guillaume. Edited and Translated by Philip E. Bennett. Grant & Cutler LTD, 2000.
La Chanson de Roland. Edited by T.A. Atkinson Jenkins. D.C. Heath and Company, 1924.
La Chanson de Roland. Edited by Joseph Bédier. Vol 2. L'edition d'Art H. Piazza, 1968.
La Chanson de Roland. Edited by Gérard Moignet. Bibliothèque Bordas, 1969.
Le Charroit de Nîmes. Translated by Fabienne Gégou. Editions Champion, 1984.
Liebermann, Felix, editor. *Die Gesetze der Angelsachsen*. Vol. 1. S.M. Niemeyer, 1916.
Lévi-Strauss, Claude. *Structural Anthropology*. Translated by Claire Jacobson and Brook Grundfest Schoepf, Basic Books, 1963.
Lidell, Henry G. and Scott, Robert. *A Greek-English Lexicon*. Oxford, 1968.
Lloyd-Jones, Hugh. *The Justice of Zeus*. U of California P, 1983.

Lord, Albert B. *The Singer of Tales*. Atheneum, 1976.

Matos-Ayala, S. "Chivalry in Literature and Practice: A Study of the Medieval Code of Arms before and after the Arthurian Legends and its Practice in the Battlefield." 2013.Theses and Dissertations (All). 1171.http://knowledge.library.iup.edu/etd/1171

Menéndez Pidal, R. editor. *Cantar de Mio Cid*. Translated by Alfonso Reyes. Espasa Calpe, 1991.

-------- *Castilla, la tradición y el idioma*. Espasa Calpe, 1945.

-------- *En torno al Poema del Cid*. EDHASA, 1970.

-------- *La España del Cid*. Vol. I. Editorial Plutarco, 1929.

Menéndez Pidal, Ramón. Translated by I.M. Cluzel. *La Chanson de Roland et la tradition épique des Francs*. Editions A. et J. Picard, 1960.

Michael, I., editor *Poema de Mio Cid*. Editorial Castalia, 1976.

Michelet, Fabienne L. ' Hospitality, Hostility, and Peacemaking in *Beowulf*." *Philological Quarterly*, vol 94, nos. 1/2, 2015, pp. 23-50.

Mossé, Fernand. *Manuel de la langue gotique*. Aubier, 1956.

Niles, J.D. "Ring Composition and the Structure of Beowulf." PMLA, vol 94, no. 4, 1974, pp. 924-935.

Orchard, Andy. *A Critical Companion to "Beowulf."* D.S. Brewer, 2007.

Owen-Crocker, Gale R. *The Four Funerals in Beowulf*. Manchester UP, 2009.

Owen, D.D.R. "Charlemagne's Dreams, Baligant and Turoldus." *Zeitschrift für Romanische Philologie* 87, 1971, pp. 197-208.

Paris, Gaston. *Histoire Poétique de Charlemagne*. 1905. Slatkine Reprints, 1974.

Pérez, Rolando. "The Cid: War Machine, Self-Made Noble and Mythic Hero of Menéndez Pidal's *Nationalistic Epic.*" *Heroes and Antiheroes: A Celebration of the Cid*, edited by Anthony J. Cárdenas-Rotunno. Hispanic Society of America, 2013, pp. 125-145.

Pope, John C., editor. *Seven Old English Poems*. Bobbs-Merrill, 1976.

Renoir, Alain. "The Heroic Oath in *Beowulf*, the *Chanson de Roland*, And the *Nibelungenlied.*" *Studies in Old English Literature in Honor of Arthur G. Brodeur*, edited by Stanley B. Greenfield. U of Oregon Books, 1963, pp. 237-267.

Robinson, Fred C. *"Beowulf" and the Appositive Style*. U of Tennessee P. 1985.

-------- *The Tomb of Beowulf and Other Essays on Old English*. Blackwell Publishers, 1993.

Robinson, Marsha R. *Matriarchy, Patriarchy, and Imperial Security in Africa*. Lexington Books, 2012.

Rychner, Jean. *La chanson de geste, essai sur l'art epique des jongleurs*. Genève: Librairie E. Droz, 1955.

Schaefer, Ursula. "Rhetoric and Style." *A Beowulf Handbook,* edited by Robert E. Bjork and John D. Niles. U of Nebraska P, 1997, pp. 105-124.

Semple, Benjamin M. ' Recognizing Roland: The Response of the Medieval Audience to the Dreams of Charlemagne in the *Song of Roland*." In *Dreams in French Literature: The Persistent Voice*. Edited by Tom Conner. Editions Rodopi, 1995.

Shippey, Thomas A. "Structure and Unity." *A Beowulf Handbook,* edited by Robert E. Bjork and John D. Niles. U of Nebraska P, 1997, pp. 149-174.

Short, I., editor. *La Chanson de Roland*. Librairie Générale Française, 1990.

Simonetti, S. How Jesus Christ Upended the Scapegoat Myth: A Girardian Interpretation. April 18, 2019. blog.action.org/archives/108119-how-jesus-christ-upended-the-scapegoat-myth-a-girardian-interpretation.html

Smyth, Herbert Weir. *Greek Grammar*. Benediction Classics, 2010.

Spitzer, L. "Sobre el carácter histórico del *Cantar de Mio Cid*." *NRFH*, 2, 1948, pp. 105-117.

Steinmeyer, Karl-Josef. *Untersuchungen zur allegorischen Bedeutung der Traüme im altfranzosischen Rolandslied*. Max Hueber Verlag, 1963.

Straczek, B. "René Girard's Concept of Mimetic Desire, Scapegoat Mechanism and Biblical Demystification." Poszukiwania Naukowe. 35, 2014, pp. 47-56.

10.21852/sem.2014.4.04

The Companion Bible. Companion Bible: The Authorized Version of 1611, with Structures and Critical, Explanatory, and Suggestive Notes and with 198 Appendixes. Edited by E.W. Bullinger. Kregel Publications, 2000.

The Free Encyclopedia. http://en.wikipedia.org/wiki/Almace

The Iliad of Homer. Translated by R. Lattimore. The U of Chicago P, 1961

The New Scofield Study Bible: New King James Version. Edited by C.I. Scofield. 1967. Consulting editor Arthur L. Farstad. Thomas Nelson Publisher, 1989.

The Nibelungenlied. Translated by Cyril Edwards. Oxford U P, 2010.

The Odyssey of Homer. Translated by R. Lattimore. Harper Perennial, 1991.

The Poem of the Cid. Translated by R. Hamilton and J. Perry. Edited by Ian Michael. Penguin Books, 1984.

The Song of Roland. Translated by Patricia Terry. Bobbs-Merrill, 1979.

Van Emden, W.G. "Another Look at Charlemagne's Dreams in the *Chanson de Roland*." *French Studies* 27, July 1974, pp. 257-270.

Voegelin, Eric. *Order in History*. Vol. 2. *The World of the Polis*. Louisiana State U, 1957.

Webster, Noah. *Webster's New Twentieth Century Dictionary of the English Language*. Edited by Jean L. McHechnie. Vol 1-2. The World Publishing Co., 1965.

Whitehead, Frederick. "Charlemagne's Second Dream." *Olifant. A Publication of the Société Rencesvals. American-Canadian Branch* 3, March 1976, pp. 189-195.

Williams, David. *Cain and "Beowulf:" A Study in Secular Allegory*. U of Toronto P, 1982.

Zahareas, A. "The Cid's Legal Action at the Court of Toledo. *RR*, 56, 1965, pp. 161-172.

www.ingramcontent.com/pod-product-compliance
Lightning Source LLC
Chambersburg PA
CBHW061322220326
41599CB00026B/4988